IDIOT'S GUIDES.
AS EASY AS IT GETS!

Calculus I

by W. Michael Kelley

ALPHA
A member of Penguin Random House LLC

Publisher: Mike Sanders
Associate Publisher: Billy Fields
Acquisitions Editor: Jan Lynn
Development Editor: Mike Thomas
Cover Designer: Lindsay Dobbs
Book Designer: William Thomas
Compositors: Ayanna Lacey, Brian Massey
Proofreader: Virginia Vasquez
Indexer: Heather McNeill

First American Edition, 2016
Published in the United States by DK Publishing
6081 E. 82nd Street, Indianapolis, Indiana 46250

Published in the United States by Dorling Kindersley Limited.

Formerly published as *The Complete Idiot's Guide to Calculus, Second Edition.*

IDIOT'S GUIDES and Design are trademarks of Penguin Random House LLC

ISBN: 9781465451682
Library of Congress Catalog Card Number: 2015956727

Note: This publication contains the opinions and ideas of its author(s). It is intended
to provide helpful and informative material on the subject matter covered. It is sold
with the understanding that the author(s) and publisher are not engaged in rendering
professional services in the book. If the reader requires personal assistance or
advice, a competent professional should be consulted. The author(s) and publisher
specifically disclaim any responsibility for any liability, loss, or risk, personal or
otherwise, which is incurred as a consequence, directly or indirectly, of the use and
application of any of the contents of this book.

DK books are available at special discounts when purchased in bulk for sales
promotions, premiums, fund-raising, or educational use. For details, contact: DK
Publishing Special Markets, 345 Hudson Street, New York, New York 10014 or
SpecialSales@dk.com.

Printed and bound in the United States of America

idiotsguides.com

Contents

Foreword

Here's a new one—a calculus book that doesn't take itself too seriously! I can honestly say that in all my years as a math major, I've never come across a book like this.

My name is Danica McKellar. I am primarily an actress and filmmaker (probably most recognized by my role as "Winnie Cooper" on *The Wonder Years*), but a while back I took a 4-year sidetrack and majored in Mathematics at UCLA. During that time I also co-authored the proof of a new math theorem and became a published mathematician. What can I say? I love math!

But let's face it. You're not buying this book because you love math. And that's okay. Frankly, most people don't love math as much as I do … or at all, for that matter. This book is not for the dedicated math majors who want every last technical aspect of each concept explained to them in precise detail.

This book is for every Bio major who has to pass two semesters of calculus to satisfy the university's requirements. Or for every student who has avoided mathematical formulas like the plague, but is suddenly presented with a whole textbook full of them. I knew a student who switched majors from chemistry to English, in order to avoid calculus!

Mr. Kelley provides explanations that give you the broad strokes of calculus concepts—and then he follows up with specific tools (and tricks!) to solve some of the everyday problems that you will encounter in your calculus classes.

You can breathe a sigh of relief—the content of this book will not demand of you what your other calculus textbooks do. I found the explanations in this book to be, by and large, friendly and casual. The definitions don't concern themselves with high-end accuracy, but will bring home the essence of what the heck your textbook was trying to describe with their 50-cent math words. In fact, don't think of this as a textbook at all. What you will find here is a conversation on paper that will hold your hand, make jokes(!), and introduce you to the major topics you'll be required to learn for your current calculus class. The friendly tone of this book is a welcome break from the clinical nature of every other math book I've ever read!

And oh, Mr. Kelley's colorful metaphors—comparing piecewise functions to Frankenstein's body parts—well, you'll understand when you get there.

My advice would be to read the chapters of this book as a nonthreatening introduction to the basic calculus concepts, and then for fine-tuning, revisit your class's textbook. Your textbook explanations should make much more sense after reading this book, and you'll be more confident and much better qualified to appreciate the specific details required of you by your class. Then you can remain in control of how detailed and nitpicky you want to be in terms of the mathematical precision of your understanding by consulting your "unfriendly" calculus textbook.

Congratulations for taking on the noble pursuit of calculus! And even more congratulations to you for being proactive and buying this book. As a supplement to your more rigorous textbook, you won't find a friendlier companion.

Good luck!

Danica McKellar
Actress, summa cum laude, Bachelor of Science in Pure Mathematics at UCLA

Introduction

Let's be honest. Most people would like to learn calculus as much as they'd like to be kicked in the face by a mule. Usually, they have to take the course because it's required or they walked too close to the mule, in that order. Calculus is dull, calculus is boring, and calculus didn't even get you anything for your birthday.

It's not like you didn't try to understand calculus. You even got this bright idea to try and read your calculus textbook. What a joke that was. You're more likely to receive the Nobel Prize for chemistry than to understand a single word of it. Maybe you even asked a friend of yours to help you, and talking to her was like trying to communicate with an Australian aborigine. You guys just didn't speak the same language.

You wish someone would explain things to you in a language that you understand, but in the back of your mind, you know that the math lingo is going to come back to haunt you. You're going to have to understand it in order to pass this course, and you don't think you've got it in you. Guess what? You do!

Here's the thing about calculus: things are never as bad as they seem. The mule didn't mean it, and I know this great plastic surgeon. I also know how terrifying calculus is. The only thing scarier than learning it is teaching it to 35 high school students in a hot, crowded room right before lunch. I've fought in the trenches at the front line and survived to tell the tale. I can even tell it in a way that may intrigue, entertain, and teach you something along the way.

We're going to journey together for a while. Allow me to be your guide in the wilderness that is calculus. I've been here before and I know the way around. My goal is to teach you all you'll need to know to survive out here on your own. I'll explain everything in plain and understandable English. Whenever I work out a problem, I'll show you every step (even the simple ones) and I'll tell you exactly what I'm doing and why. Then you'll get a chance to practice the skill on your own without my guidance. Never fear, though—I answer the question for you fully and completely in the back of the book.

I'm not going to lie to you. You're not going to find every single problem easy, but you will eventually do every one. All you need is a little push in the right direction, and someone who knows how you feel. With all these things in place, you'll have no trouble hoofing it out. Oh, sorry, that's a bad choice of words.

How This Book Is Organized

This book is presented in five parts.

In **Part 1, The Roots of Calculus,** you'll learn why calculus is useful and what sorts of skills it adds to your mathematical repertoire. You'll also get a taste of its history, which is marred by quite a bit of controversy. Being a math person, and by no means a history buff, I'll get into the math without much delay. However, before we can actually start discussing calculus concepts, we'll spend some quality time reviewing some prerequisite algebra and trigonometry skills.

In **Part 2, Laying the Foundation for Calculus,** it's time to get down and dirty. This is the moment you've been waiting for. Or is it? Most people consider calculus the study of derivatives and integrals, and we don't really talk too much about those two guys until Part 3. Am I just a royal tease? Nah. First, we have to talk about limits and continuity. These foundational concepts constitute the backbone for the rest of calculus, and without them, derivatives and integrals couldn't exist.

Finally, we meet one of the major players in **Part 3, The Derivative.** The name says it all. All of your major questions will be answered, including what a derivative is, how to find one, and what to do if you run into one in a dark alley late at night. (Run!) You'll also learn a whole slew of major derivative-based skills: drawing graphs of functions you've never seen, calculating how quickly variables change in given functions, and finding limits that once were next to impossible to calculate. But wait, there's more! How could something called a "wiggle graph" be anything but a barrel of giggles?

In **Part 4, The Integral,** you meet the other big boy of calculus. Integration is almost the same as differentiation, except that you do it backward. Intrigued? You'll learn how the area underneath a function is related to this backward derivative, called an "antiderivative." It's also time to introduce the Fundamental Theorem of Calculus, which (once and for all) describes how all this crazy stuff is related. You'll find out that integrals are a little more disagreeable than derivatives were; they require you to learn more techniques, some of which are extremely interesting and (is it possible?) even a little fun!

Now that you've met the leading actor and actress in this mathematical drama, what could possibly be left? The love story, of course! In **Part 5, Differential Equations,** I weave a beautiful narrative detailing the intricate relationship between derivatives and integrals sharing their lives together in a small, rural suburban neighborhood. Well, that's not entirely true, but you do get to play with fun things called slope fields and you end this part by taking an exam on all the content in the book and get even more practice! What could be better than that?

At the back of the book, I've included the solutions to all the practice problems as well as a glossary of helpful terms.

Extras

As a teacher, I constantly found myself going off on tangents—everything I mentioned reminded me of something else. These peripheral snippets are captured in this book as well. Here's a guide to the different sidebars you'll see peppering the pages that follow.

CRITICAL POINT

These notes, tips, and thoughts will assist, teach, and entertain. They add a little something to the topic at hand, whether it be sound advice, a bit of wisdom, or just something to lighten the mood a bit.

DEFINITION

Calculus is chock-full of crazy- and nerdy-sounding words and phrases. In order to become King or Queen Math Nerd, you'll have to know what they mean!

KELLEY'S CAUTIONS

Although I will warn you about common pitfalls and dangers throughout the book, the dangers in these boxes deserve special attention. Think of these as skulls and crossbones painted on little signs that stand along your path. Heeding these cautions can sometimes save you hours of frustration.

YOU'VE GOT PROBLEMS

Math is not a spectator sport! After we discuss a topic, I'll explain how to work out a certain type of problem, and then you have to try it on your own. These problems will be very similar to those that I walk you through in the chapters, but now it's your turn to shine. Even though all the answers appear in Appendix A, you should only look there to check your work.

Dedication

This book is dedicated to Lisa, who is no Linda Ronstadt. Despite not knowing much, I know I love you. I also know that one day our hit single about ham that we dropped on the floor will be a worldwide phenomenon.

To my kids Nick, Erin, and Sara. I know that I will miss the noise (God in heaven, the noise) of you playing while I am working. One day soon you will be old enough to have loud children of your own. I love you very much. Now please shush—Dad is trying to write.

Finally, to Joe. It was a home run.

Special Thanks to the Technical Reviewers

Idiot's Guides: Calculus I was reviewed by Robert Halstead, an expert who double-checked the accuracy of what you'll learn here. He's also the kind of super nice guy who helps you move furniture even when his shoulder is hanging out of its socket. The publisher would like to extend our thanks to Rob for helping us ensure that this book gets all its facts straight. We also thank Sue Strickland, who reviewed the previous editions and is still the best mathematics instructor that ever was.

Rob is a mathematics teacher at Northern High School in Calvert County, Maryland, with 22 years of teaching experience. He spent the last 15 of those years teaching Advanced Placement Calculus. He has served as the Mathematics Core Lead and department chair at his school, and he was chosen as the school's Teacher of the Year in 2012.

Susan received a BS in Mathematics from St. Mary's College of Maryland in 1979, an MS in Mathematics from Lehigh University in Bethlehem, Pennsylvania, in 1982, and took graduate courses in Mathematics Education at The American University in Washington, D.C., from 1989 through 1991. She was an assistant professor of mathematics and supervised student mathematics teachers at St. Mary's College of Maryland from 1983 through 2001. In the summer of 2001, she accepted the position as a professor of mathematics at the College of Southern Maryland, where she expects to be until she retires! Her interests include teaching mathematics to the "math phobics," training new math teachers, and solving math games and puzzles.

The Roots of Calculus

You've heard of Newton, haven't you? If not the man, then at least the fruit-filled cookie? Well, the Sir Isaac variety of Newton is one of the two men responsible for bringing calculus into your life and your course-requirement list (or maybe I should say, one of the two men who should shoulder the blame). Calculus's history is long, however, and its concepts predate either man. Before we start studying calculus, we'll take a (very brief) look at its history and development and answer that sticky question: "Why do I have to learn this?"

Next, it's off to practice our prerequisite math skills. You wouldn't try to bench-press 300 pounds without warming up first, would you? A quick review of linear equations, factoring, quadratic equations, function properties, and trigonometry will do a body good. Even if you think you're ready to jump right into calculus, this brief review is recommended. I bet you've forgotten a few things you'll need to know later, so take care of that now!

What Is Calculus, Anyway?

The word *calculus* can mean one of two things: a computational method or a mineral growth in a hollow organ of the body, such as a kidney stone. Either definition personifies the pain and anguish often endured by students trying to understand the subject. It is far from controversial to suggest that mathematics is not the most popular of subjects in contemporary education; in fact, calculus holds the great distinction of King of the Evil Math Realm, especially by the math phobic. It represents an unattainable goal, an unthinkable miasma of confusion and complication, and few venture into its realm unless propelled by such forces as job advancement or degree requirement. No one knows how much people fear calculus more than a calculus teacher.

The minute people find out I taught a calculus class, they are compelled to describe, in great detail, exactly how they did in high school math, what subject they "topped out" in, and why they feel that calculus is the embodiment of evil. Most of these people are my barbers, and I can't explain why. All of the friendly folks at the Hair Cuttery have come to know me as the strange balding man with arcane and baffling mathematical knowledge.

In This Chapter

- Why calculus is useful
- The historic origins of calculus
- The authorship controversy
- Can I ever learn this?

Most of the fears surrounding calculus are unjustified. Calculus is a step up from high school algebra, no more. Following a straightforward list of steps, just like you do with most algebra problems, solves the majority of calculus problems. Don't get me wrong—calculus is not always easy, and the problems are not always trivial, but it is not as imposing as it seems. Calculus is a truly fascinating tool with innumerable applications to "real life," and for those of you who like soap operas, it's got one of the biggest controversies in history to its credit.

> ✏️ **CRITICAL POINT**
>
> What we call "calculus," scholars call "*the* calculus." Because any method of computation can be called a calculus and the discoveries comprising modern-day calculus are so important, the distinction is made to clarify. I personally find the terminology a little pretentious and won't use it. I've never been asked, "Which calculus are you talking about?"

What's the Purpose of Calculus?

Calculus is a very versatile and useful tool, not a one-trick pony by any stretch of the imagination. Many of its applications are direct upgrades from the world of algebra—methods of accomplishing similar goals, but in a far greater number of situations. Whereas it would be impossible to list all the uses of calculus, the following list represents some interesting highlights of the things you will learn by the end of the book.

Finding the Slopes of Curves

One of the earliest algebra topics learned is how to find the slope of a line—a numerical value that describes just how slanted that line is. Calculus affords us a much more generalized method of finding slopes. With it, we can find not only how steeply a line slopes, but indeed, how steeply any curve slopes at any given time. This might not at first seem useful, but it is actually one of the most handy mathematics applications around.

Calculating the Area of Bizarre Shapes

Without calculus, it is difficult to find areas of shapes other than those whose formulas you learned in geometry. Sure, you may be a pro at finding the area of a circle, square, rectangle, or triangle, but how would you find the area of a shape like the one shown in Figure 1.1?

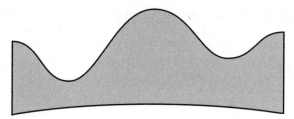

Figure 1.1
Calculate this area? We're certainly not in Kansas anymore ....

Justifying Old Formulas

There was a time in your math career when you took formulas on faith. Sometimes we still need to do that, but calculus affords us the opportunity to finally verify some of those old formulas, especially from geometry. You were always told that the volume of a cone was one-third the volume of a cylinder with the same radius $\left(V = \frac{1}{3}\pi r^2 h\right)$, but through a simple calculus process of three-dimensional linear rotation, we can finally prove it. (By the way, the process really is simple even though it may not sound like it right now.)

Calculating Complicated x-Intercepts

Without the aid of a graphing calculator, it is exceptionally hard to calculate an *irrational root*. However, a simple, repetitive process called Newton's Method (named after Sir Isaac Newton) enables you to calculate an irrational root to whatever degree of accuracy you desire.

DEFINITION

An **irrational root** is an x-intercept that is not a fraction. Fractional (rational) roots are much easier to find, because you can typically factor the expression to calculate them, a process that is taught in the earliest algebra classes. No good, generic process of finding irrational roots is possible until you use calculus.

Visualizing Graphs

You may already have a good grasp of lines and how to visualize their graphs easily, but what about the graph of something like $y = x^3 + 2x^2 - x + 1$? Very elementary calculus tells you exactly where that graph will be increasing, decreasing, and twisting. In fact, you can find the highest and lowest points on the graph without plotting a single point.

Finding the Average Value of a Function

Anyone can average a set of numbers, given the time and the fervent desire to divide. Calculus enables you to take your averaging skills to an entirely new level. Now you can even find, on average, what height a function travels over a period of time. For example, if you graph the path of an airplane (see Figure 1.2), you can calculate its average cruising altitude with little or no effort. Determining its average velocity and acceleration are no harder. You may never have had the impetus to do such a thing, but you've got to admit that it's certainly more interesting than averaging the odd numbers less than 50.

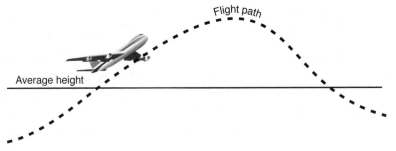

Figure 1.2
Even though this plane's flight path is not defined by a simple shape (like a semicircle), using calculus you can calculate all sorts of things, like its average altitude during the journey or the number of complimentary peanuts you dropped when you fell asleep.

Calculating Optimal Values

One of the most mind-bendingly useful applications of calculus is the optimization of functions. In just a few steps, you can answer questions such as "If I have 1,000 feet of fence, what is the largest rectangular yard I can make?" or "Given a rectangular sheet of paper which measures 8.5 inches by 11 inches, what are the dimensions of the box I can make containing the greatest volume?" The traditional way to create an open box from a rectangular surface is to cut congruent squares from the corners of the rectangle and then to fold the resulting sides up, as shown in Figure 1.3.

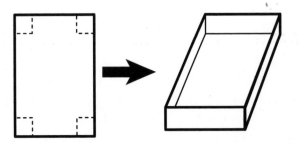

Figure 1.3
With a few folds and cuts, you can easily create an open box from a regular surface.

I tend to think of learning calculus and all of its applications as suddenly growing a third arm. Sure, it may feel funny having a third arm at first. In fact, it'll probably make you stand out in bizarre ways from those around you. However, given time, you're sure to find many uses for that arm that you'd have never imagined without having first possessed it.

Who's Responsible for This?

Tracking the discovery of calculus is not as easy as, say, tracking the discovery of the safety pin. Any new mathematical concept is usually the result of hundreds of years of investigation, debate, and debacle. Many come close to stumbling upon key concepts, but only the lucky few who finally make the small, key connections receive the credit. Such is the case with calculus.

Calculus is usually defined as the combination of the differential and integral techniques you will learn later in the book. However, historical mathematicians would never have swallowed the concepts we take for granted today. The key ingredient missing in mathematical antiquity was the hairy notion of infinity. Mathematicians and philosophers of the time had an extremely hard time conceptualizing infinitely small or large quantities. Take, for instance, the Greek philosopher Zeno.

Ancient Influences

Zeno took a very controversial position in mathematical philosophy: he argued that all motion is impossible. In the paradox titled Dichotomy, he used a compelling, if not strange, argument illustrated in Figure 1.4.

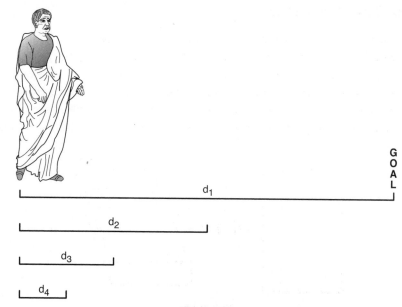

Figure 1.4

The infinite subdivisions described in Zeno's Dichotomy.

> **✏ CRITICAL POINT**
>
> The most famous of Zeno's paradoxes is a race between a tortoise and the legendary Achilles called, appropriately, Achilles and the Tortoise. Zeno contends that if the tortoise has a head start, no matter how small, Achilles will never be able to close the distance. To do so, he'd have to travel half of the distance separating them, then half of that, *ad nauseum,* presenting the same dilemma illustrated by the Dichotomy paradox.

In Zeno's argument, the individual pictured wants to travel to the right, to his eventual destination. However, before he can travel that distance (d_1), he must first travel half of that distance (d_2). That makes sense, since d_2 is smaller and comes first in the path. However, before the d_2 distance can be completed, he must first travel half of it (d_3). This procedure can be repeated indefinitely, which means that our beleaguered sojourner must travel an infinite number of distances. No one can possibly do an infinite number of things in a finite amount of time, says Zeno, since an infinite list will never be exhausted. Therefore, not only will the man never reach his destination, he will, in fact, never start moving at all! This could account for the fact that you never seem to get anything done on Friday afternoons.

Zeno didn't actually believe that motion was impossible. He just enjoyed challenging the theories of his contemporaries. What he, and the Greeks of his time, lacked was a good understanding of infinite behavior. It was unfathomable that an innumerable number of things could fit into a measured, fixed space. Today, geometry students accept that a line segment, though possessing fixed length, contains an infinite number of points. The development of a reasonable and yet mathematically sound concept of very large quantities or very small quantities was required before calculus could sprout.

> **✏ CRITICAL POINT**
>
> In case the suspense is killing you, let me ruin the ending for you. The essential link to completing calculus and satisfying everyone's concerns about infinite behavior was the concept of limit, which laid the foundation for both derivatives and integrals.

Some ancient mathematicians weren't troubled by the apparent contradiction of an infinite amount in a finite space. Most notably, Euclid and Archimedes contrived the method of exhaustion as a technique to find the area of a circle, since the exact value of π wouldn't be around for some time. In this technique, regular polygons were inscribed in a circle; the higher the number of sides of the polygon, the closer the area of the polygon came to the area of the circle (see Figure 1.5).

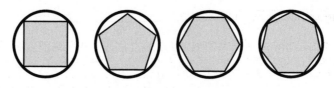

Figure 1.5

The higher the number of sides, the closer the area of the inscribed polygon comes to approximating the area of the circle.

In order for the method of exhaustion (which is aptly titled, in my opinion) to give the exact value for the circle, the polygon would have to have an infinite number of sides. Indeed, this magical incarnation of geometry can only be considered theoretically, and the idea that a shape of infinite sides could have a finite area made most people of the time very antsy. However, seasoned calculus students of today can see this as a simple limit problem. As the number of sides approaches infinity, the area of the polygon approaches πr^2, where r is the radius of the circle. Limits are essential to the development of both the derivative and integral, the two fundamental components of calculus.

Although Newton and Leibniz were unearthing the major discoveries of calculus in the late 1600s and early 1700s, no one had established a formal limit definition. Although this may not keep *us* up at night, it was, at the least, troubling at the time. Mathematicians worldwide started sleeping more soundly at night circa 1751, when Jean Le Rond d'Alembert wrote *Encyclopédie* and established the formal definition of the limit. The delta-epsilon definition of the limit we use today is very close to that of d'Alembert.

Even before its definition was established, however, Newton had given a good enough shot at it that calculus was already taking shape.

Newton vs. Leibniz

Sir Isaac Newton, who was born in poor health in 1642 but became a world-renowned smart guy (even during his own time), once retorted, "If I have seen farther than Descartes, it is because I have stood on the shoulders of giants." No truer thing could be said about any major mathematical discovery, but let's not give the guy too much credit for his supposed modesty (more to come on that in a bit). Newton realized that infinite series (e.g., the method of exhaustion) were not only great approximators, but if allowed to actually reach infinity, they gave the exact values of the functions they approximated. Therefore, they behaved according to easily definable laws and restrictions usually only applied to known functions. Most importantly, he was the first person to recognize and utilize the inverse relationship between the slope of a curve and the area beneath it.

That inverse relationship (contemporarily called the Fundamental Theorem of Calculus) marks Newton as the inventor of calculus. He published his findings, and his intuitive definition of a limit, in his 1687 masterwork entitled *Philosophiæ Naturalis Principia Mathematica*. The *Principia*, as it is more commonly known today, is considered by some (those who consider such things, I suppose) to be the greatest scientific work of all time, excepting of course any books yet to be written by the comedian Sinbad. Calculus was actively used to solve the major scientific dilemmas of the time:

- Calculating the slope of the tangent line to a curve at any point along its length

- Determining the velocity and acceleration of an object given a function describing its position, and designing such a position function given the object's velocity or acceleration

- Calculating arc lengths and the volume and surface area of solids

- Calculating the relative and absolute *extrema* of objects, especially projectiles

DEFINITION

Extrema points are high or low points of a curve (maxima or minima, respectively). In other words, they represent extreme values of the graph, whether extremely high or extremely low, in relation to the points surrounding them.

However, with a great discovery often comes great controversy, and such is the case with calculus.

Enter Gottfried Wilhelm Leibniz, child prodigy and mathematical genius. Leibniz was born in 1646 and completed college, earning his Bachelor's degree, at the ripe old age of 17. Because Leibniz was primarily self-taught in the field of mathematics, he often discovered important mathematical concepts on his own, long after someone else had already published them. Newton actually credited Leibniz in his *Principia* for developing a method similar to his. That similar method evolved into a near match of Newton's work in calculus, and in fact, Leibniz published his breakthrough work inventing calculus *before* Newton, although Newton had already made the exact discovery years before Leibniz. Some argue that Newton possessed extreme sensitivity to criticism and was, therefore, slow to publish. The mathematical war was on: who invented calculus first and thus deserved the credit for solving a riddle thousands of years old?

CRITICAL POINT

Ten years after Leibniz's death, Newton erased the reference to Leibniz from the third edition of the *Principia* as a final insult. This is approximately the academic equivalent of Newton throwing a chair at Leibniz on *The Jerry Springer Show* (topic: "You published your solution to an ancient mathematical riddle before me and I'm fightin' mad!").

Today, Newton is credited for inventing calculus first, although Leibniz is credited for its first publication. In addition, the shadow of plagiarism and doubt has been lifted from Leibniz, and it is believed that he discovered calculus completely independent of Newton. However, two distinct factions arose and fought a bitter war of words. British mathematicians sided with Newton, whereas continental Europe supported Leibniz, and the war was long and hard. In fact, British mathematicians were effectively alienated from the rest of the European mathematical community because of the rift, which probably accounts for the fact that there were no great mathematical discoveries made in Britain for some time thereafter.

Although Leibniz just missed out on the discovery of calculus, many of his contributions live on in the language and symbols of mathematics. In algebra, he was the first to use a dot to indicate multiplication ($3 \cdot 4 = 12$) and a colon to designate a proportion ($1{:}2 = 3{:}6$). In geometry, he contributed the symbols for congruent ($\cong$) and similar ($\sim$). Most famous of all, however, are the symbols for the derivative and the integral, which we also use.

Will I Ever Learn This?

History aside, calculus is an overwhelming topic to approach from a student's perspective. There are an incredible number of topics, some of which are related, but most of which are not in any obvious sense. However, there is no topic in calculus that is, in and of itself, very difficult once you understand what is expected of you. The real trick is to quickly recognize what sort of problem is being presented and then to attack it using the methods you will read and learn in this book.

CRITICAL POINT

Leibniz also coined the term *function,* which is commonly learned in an elementary algebra class. However, most of Leibniz's discoveries and innovations were eclipsed by Newton, who made great strides in the topics of gravity, motion, and optics (among other things). The two men were bitter rivals and were fiercely competitive against each other.

I have taught calculus for a number of years, to high school students and adults alike, and I believe that there are four basic steps to succeeding in calculus:

Make sure to understand what the major vocabulary words mean. This book will present all important vocabulary terms in simple English, so you understand not only what the terms mean, but how they apply to the rest of your knowledge.

Sift through the complicated wording of the important calculus theorems and strip away the difficult language. Math is just as foreign a language as French or Spanish to someone who doesn't enjoy numbers, but that doesn't mean you can't understand complicated mathematical theorems. I will translate every theorem into plain English and make all the underlying implications perfectly clear.

Develop a mathematical instinct. As you read, I will help you recognize subtle clues presented by calculus problems. Most problems do everything but tell you exactly how they must be solved. If you read carefully, you will develop an instinct, a feeling that will tingle in your inner fiber and guide you toward the right answers. This comes with practice, practice, practice, so I'll provide sample problems with detailed solutions to help you navigate the muddy waters of calculus.

Sometimes you just have to memorize. There are some very advanced topics covered in calculus that are hard to prove. In fact, many theorems cannot be proven until you take much more advanced math courses. Whenever I think that proving a theorem will help you understand it better, I will do so and discuss it in detail. However, if a formula, rule, or theorem has a proof that I deem unimportant to your mastering the topic in question, I will omit it, and you'll just have to trust me that it's for the best.

The Least You Need to Know

- Calculus is the culmination of algebra and geometry.
- Calculus as a tool enables us to achieve greater feats than the mathematics courses that precede it.
- Limits are foundational to calculus.
- Newton and Leibniz both discovered calculus independently, though Newton discovered it first.
- With time and dedication, anyone can be a successful calculus student.

Polish Up Your Algebra Skills

If you are an aspiring calculus student, somewhere in your past you probably had to do battle with the beast called algebra. Not many people have positive memories associated with their algebraic experiences, and I am no different. Forget the fact that I was a math major, a calculus teacher, and even took my calculator to bed with me when I was young (a true but very sad story). I hated algebra for many reasons, not the least of which was that I felt I could never keep up with it. Every time I seemed to understand algebra, we'd be moving on to a new topic much harder than the last.

Being an algebra student is sort of like battling a famous boxer. Here is this champion of mathematical reasoning that has stood unchallenged for hundreds of years, and you're in the ring going toe-to-toe with it. You never really reach back for that knockout punch because you're too busy fending off your opponent's blows. When the bell rings to signal the end of the fight, all you can think is "I survived!" and hope that someone can carry you out of the ring.

Perhaps you didn't hate algebra as much as I did. You might be one of those lucky people who understood algebra easily. You are very lucky. For the rest of us, however, there is hope. Algebra is much easier in retrospect than when you were

In This Chapter

* Creating linear equations
* The properties of exponents
* Factoring polynomials
* Solving quadratic equations

first being pummeled by it. As calculus is a grand extension of algebra, you will, of course, need a large repertoire of algebra skills. So it's time to slip those old boxing gloves back on and go a few rounds with your old sparring partner. The good news is you've undoubtedly gotten stronger since the last bout. If, however, a brief algebra review is not enough for you, pick up this book's prequel, *The Complete Idiot's Guide to Algebra,* by yours truly.

Walk the Line: Linear Equations

Graphs play a large role in calculus, and the simplest of graphs, the line, surprisingly pops up all the time. As such, it is important that you can recognize, write, and analyze graphs and equations of lines. To begin, remember that a line's equation always has three components: two variable terms and a constant (numeric) term. One of the most common ways to write an equation is in standard form.

Common Forms of Linear Equations

A line in standard form looks like this: $Ax + By = C$. In other words, the variable terms are on the left side and the number is on the right side of the equal sign. Also, to officially be in standard form, the coefficients (A, B, and C) must be *integers*, and A is supposed to be positive. What's the purpose of standard form? A linear equation can have many different forms (for example, $x + y = 2$ is the same line as $x = 2 - y$). However, once in standard form, all lines with the same graph have the exact same equation. Therefore, standard form is especially handy for instructors; they'll often ask that answers be put into standard form to avoid alternate correct answers.

> 📖 **DEFINITION**
>
> An **integer** is a number without a decimal or fractional part. For example, 3 and –6 are integers, whereas 10.3 and $-\frac{1}{2}$ are not.

 YOU'VE GOT PROBLEMS

> Problem 1: Express in standard form:
> $$3x - 4y - 1 = 9x + 5y - 12$$

There are two major ways to create the equation of a line. One requires that you have the slope and the *y*-intercept of the line. Appropriately enough, it is called slope-intercept form: $y = mx + b$. In this equation, *m* represents the slope and *b* the *y*-intercept. Notice the major characteristic of an equation in slope-intercept form: it is solved for *y*. In other words, *y* appears by itself on the left side of the equation.

Example 1: Write the equation of a line with slope -3 and y-intercept 5.

Solution: In slope intercept form, $m = -3$ and $b = 5$, so plug those into the slope-intercept formula:

$$y = mx + b$$
$$y = -3x + 5$$

Another way to create a linear equation requires a little less information—only a point and the slope (the point doesn't have to be the y-intercept). This (thanks to the vast creativity of mathematicians) is called point-slope form. Given the point (x_1, y_1) and slope m, the equation of the resulting line will be $y - y_1 = m(x - x_1)$.

You will find this form extremely handy throughout the rest of your travels with calculus, so make sure you understand it. Don't get confused between the x's and x_1's or the y's and the y_1's. The variables with the subscript represent the coordinates of the point you're given. Don't replace the other x and y with anything—these variables are left in your final answer. Watch how easy this is.

Example 2: If a line g contains the point $(-5,2)$ and has slope $-\frac{1}{5}$, what is the equation of g in standard form?

Solution: Because you are given a slope and a point (which is not the y-intercept), you should use point-slope form to create the equation of the line. Therefore, $m = -\frac{1}{5}$, $x_1 = -5$, and $y_1 = 2$. Plug these values into point-slope form and get:

$$y - 2 = -\frac{1}{5}\left(x - (-5)\right)$$
$$y - 2 = -\frac{1}{5}\left(x + 5\right)$$

If this equation is supposed to be in standard form, you're not allowed to have any fractions. Remember that the coefficients have to be integers, so to get rid of the fractions, multiply the entire equation by 5:

$$5y - 10 = -\left(x + 5\right)$$
$$5y - 10 = -x - 5$$

Now, move the variables to the left and the constants to the right and make sure the x term is positive; this puts everything in standard form:

$$x + 5y = 5$$

YOU'VE GOT PROBLEMS

Problem 2: Find the equation of the line through point $(0,-2)$ with slope $\frac{2}{3}$ and put it in standard form.

Calculating Slope

You might have noticed that both of the ways we use to create lines absolutely require that you know the slope of the line. The slope of the line is *that* important (almost as important as wearing both shoes and a shirt if you want to buy a Slurpee at 7-Eleven). The *slope* of a line is a number that describes precisely how "slanty" that line is—the larger the value of the slope, the steeper the line. Furthermore, the sign of the slope (in most cases Capricorn) will tell you whether or not the line rises or falls as it travels.

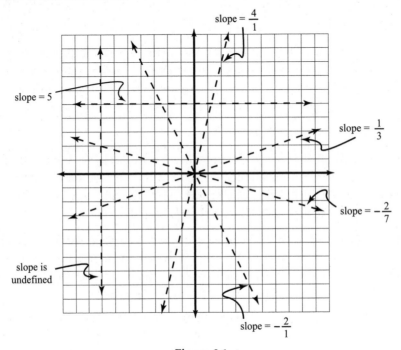

Figure 2.1
Calculating the slope of a line.

As shown in Figure 2.1, lines with shallower inclines have smaller slopes. If the line rises (from left to right), the slope is positive; if, however, it falls from left to right, the slope is negative. Horizontal lines have 0 slope (neither positive nor negative), and vertical lines are said to have an undefined slope, or no slope at all.

It is very easy to calculate the slope of any line: find any two points on the line, (a,b) and (c,d), and plug them into this formula:

$$\text{slope} = \frac{d-b}{c-a}$$

In essence, you are finding the difference in the *y*'s and dividing by the difference in the *x*'s. If the numerator is larger, the *y*'s are changing faster, and the line is getting steeper. On the other hand, if the denominator is larger, the line is moving more quickly to the left or right than up and down, creating a shallow incline.

> **YOU'VE GOT PROBLEMS**
>
> Problem 3: Find the slope of the line that contains points (3,7) and (–1,4).

You should also remember that parallel lines have equal slopes, whereas perpendicular lines have slopes that are negative reciprocals of one another. Therefore, if line *g* has slope $\frac{5}{7}$, then a parallel line *h* would have slope $\frac{5}{7}$ also; a perpendicular line *k* would have slope $-\frac{7}{5}$.

Example 3: Find the equation of line *j* given that it is parallel to the line $2x - y = 6$ and contains the point (–1,1); write *j* in slope-intercept form.

Solution: This problem requires you to create the equation of a line, and you'll find that the best way to do this every time is via point-slope form. So you need a point and a slope. Well, you already have the point: (–1,1). Using your keen sense of deduction, you know that only the slope is left to find and that'll be that. But how to find the slope?

If *j* is *parallel* to $2x - y = 6$, then the lines must have the same slope, so what's the slope of $2x - y = 6$? Here's the key: if you solve it for *y*, it will be in slope-intercept form, and the slope, *m*, is simply the coefficient of *x*. When you do so, you get $y = 2x - 6$. Therefore, the slope of both lines is 2, and you can use point-slope form to write the equation of *j*:

$$y - y_1 = m\left(x - x_1\right)$$
$$y - 1 = 2\left(x - \left(-1\right)\right)$$
$$y - 1 = 2\left(x + 1\right)$$

Solve for *y* to put the equation in slope-intercept form:

$$y - 1 = 2x + 2$$
$$y = 2x + 2 + 1$$
$$y = 2x + 3$$

Interpreting Linear Graphs

Calculus has undergone a renaissance over the last few decades, as researchers have gained more insight about the most effective way to present and learn mathematical material. Without climbing onto a soapbox, allow me to present the "bottom line": you need to learn the concepts *behind* the math, not just memorize a series of steps to reach a solution.

Understanding based on pure memorization is fragile—without constant practice, it shatters. Therefore, throughout this book (and most likely throughout your calculus course) you will be presented with nontraditional problems, including problems presented graphically. If these types of problems feel strange, don't worry. They are meant to stretch your understanding of the topic at hand, and upon wrestling with them for a bit, they provide you with a deeper and longer-lasting mastery of mathematics.

In the next example, you're not given a specific slope, intercept, or point to make a line. Instead, you're given a graph. All the information you need is in there—you just have to harvest it yourself.

Example 4: The graph of line p is presented in Figure 2.2. Express the equation of the line perpendicular to p with the same x-intercept in standard form.

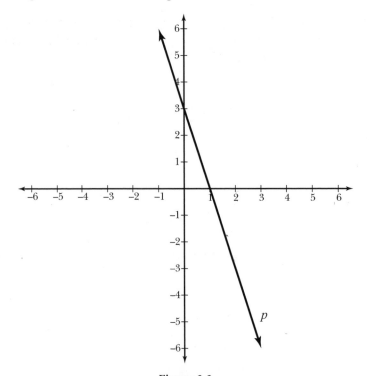

Figure 2.2
The graph of line p.

Solution: Before you jump into the solution, analyze the graph of the line and collect all the information you can, even if it later proves unnecessary. Here are my observations:

- The slope of p must be negative, because the line travels down as you move from left to right.

- The x-intercept of line p is 1, because it passes through the x-axis at point (1,0).

- The y-intercept of line p is 3, because it passes through the y-axis at point (0,3).

None of those are earth-shattering observations by any means, but that is all you need to solve the problem. You're asked to find the equation of the line perpendicular to p, which means the slope of that line is the opposite reciprocal of the slope of p. You are given two points through which p passes, so apply the slope formula with $(a,b) = (1,0)$ and $(c,d) = (0,3)$:

$$\text{slope} = \frac{d-b}{c-a}$$
$$= \frac{3-0}{0-1}$$
$$= \frac{3}{-1}$$
$$= -3$$

The slope of p is -3. My assumption that the slope was going to be negative was correct, but you didn't doubt me for a moment, did you? The slope of the line perpendicular to p must be the opposite reciprocal of -3, which is $\frac{1}{3}$.

According to the problem, the new line shares the same x-intercept as p, so the new line must also pass through point (1,0). You now know the slope of the line you are creating $\left(m = \frac{1}{3}\right)$ and a point on the line $(x_1, y_1) = (1,0)$. Apply the point-slope formula.

$$y - y_1 = m\left(x - x_1\right)$$
$$y - 0 = \frac{1}{3}\left(x - 1\right)$$
$$y = \frac{1}{3}\left(x - 1\right)$$

The line needs to be in standard form, so no fractions are allowed. Multiply both sides of the equation by 3 to eliminate the fractions.

$$3 \cdot y = \left(\frac{3}{1}\right)\left(\frac{1}{3}\right)\left(x - 1\right)$$
$$3y = 1\left(x - 1\right)$$
$$3y = x - 1$$

A linear equation in standard form has the variable terms on the left side of the equation and the constant (number term) on the right. Subtract x from both sides.

$$-x + 3y = -1$$

Almost finished! A line in standard form *must* have a positive x-coefficient, so multiply everything by -1:

$$\left(-1\right)\left(-x+3y\right)=\left(-1\right)\left(-1\right)$$
$$x - 3y = 1$$

The equation of the line perpendicular to line p that passes through the same x-intercept is $x - 3y = 1$.

 **YOU'VE GOT PROBLEMS**

Problem 4: Calculate the y-intercept of line j in Figure 2.3. *Hint: Express the equation of the line in slope-intercept form.*

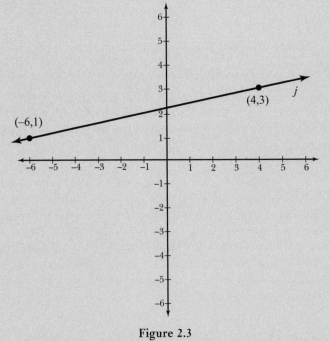

Figure 2.3

The graph of line j, including two points through which it passes.

You've Got the Power: Exponential Rules

I find that exponents are the bane of many calculus students. Whether they never learned exponents well in the first place or simply make careless mistakes, exponential errors are a treasure trove of frustration. Therefore, it's worth your while to spend a few minutes and refresh yourself on the major exponential rules. You may find this exercise empowering.

- Rule one: $x^a \cdot x^b = x^{a+b}$

Explanation: If you multiply two terms with the same base (here it's x), add the powers and keep the base. For example, $a^2 \cdot a^7 = a^9$.

- Rule two: $\dfrac{x^a}{x^b} = x^{a-b}$

Explanation: This is the opposite of rule one. If you divide (instead of multiply) two terms with the same base, then you subtract (instead of add) the powers and keep the base. For example, $\dfrac{w^7}{w^3} = w^{7-3} = w^4$.

- Rule three: $x^{-a} = \dfrac{1}{x^a}$

Explanation: A negative exponent indicates that a variable is in the wrong spot, and belongs in the opposite part of the fraction, but it only affects the variable it's touching. For example, in the expression $\frac{x^3 y^{-2}}{3}$, only the y is raised to a negative power, so it needs to be in the opposite part of the fraction. Correctly simplified, that fraction looks like this: $\frac{x^3}{3y^2}$. Note that the exponent becomes positive when it moves to the right place. Remember that a happy (positive) exponent is where it belongs in a fraction.

CRITICAL POINT

Eliminate negative exponents in your answers. Most instructors consider an answer with negative exponents in it unsimplified. They must see the glass as half-empty. Think about it. How many cheery math teachers do *you* know?

- Rule four: $(x^a)^b = x^{ab}$

Explanation: If an exponential expression is raised to a power, you should multiply the exponents and keep the base. For example, $(h^7)^3 = h^{21}$.

- Rule five: $x^{a/b} = \sqrt[b]{x^a} = \left(\sqrt[b]{x}\right)^a$

Explanation: The numerator of the fractional power remains the exponent. The denominator of the power tells you what sort of radical (square root, cube root, etc.). For example, $4^{3/2}$ can be simplified as either $\sqrt{4^3}$ or $\left(\sqrt{4}\right)^3$. Either way, the answer is 8.

Example 5: Simplify $xy^{1/3}(x^2y)^3$.

Solution: Your first step should be to raise (x^2y) to the third power. You have to use rule four twice (the current exponent of y is understood to be 1 if it is not written). This gives you $x^{2\cdot3}y^{1\cdot3} = x^6y^3$. The problem now looks like this: $xy^{1/3}\left(x^6y^3\right)$.

To finish, you have to multiply the x's and y's together using rule one:

$$x \cdot x^6 \cdot y^{1/3} \cdot y^3$$

$$= x^{1+6}y^{(1/3)+3}$$
$$= x^7 y^{10/3}$$

YOU'VE GOT PROBLEMS

Problem 5: Simplify the expression $(3x^{-3}y^2)^2$ using exponential rules.

Breaking Up Is Hard to Do: Factoring Polynomials

Factoring is one of those things you see over and over in algebra. I have found that even among my students who disliked math, factoring was popular; it's something that some people just "got," even when most everything else escaped them. This is not the case, however, in many European schools, a fact that surprised my colleagues and me when I was a high school teacher.

Canadian exchange students gave me blank stares when we discussed factoring in class. This is not to say that these students were not extremely intelligent (they were); they just used other methods. However, factoring comes in handy throughout calculus, so I deem it important enough to cover here. Call it patriotism.

Factoring is basically reverse multiplying—undoing the process of multiplication to see what was there to begin with. For example, you can break down the number 6 into factors of 3 and 2, since $3 \cdot 2 = 6$. There can be more than one correct way to factor something.

DEFINITION

Factoring is the process of "unmultiplying," breaking a number or expression down into parts that, if multiplied together, return the original quantity.

Greatest Common Factor

Use the greatest common factor method of factoring if you have terms with elements in common. It's easier than it sounds. Take the expression $4x + 8$.

Notice that both terms can be divided by 4, making 4 a common factor. Therefore, you can write the expression in the factored form of $4(x + 2)$.

In effect, I have "pulled out" the common factor of 4, and what's left behind are the terms once 4 has been divided out of each. In this type of problem, ask yourself, "What do each of the terms have in common?" and then pull that greatest common factor out of each to write your answer in factored form.

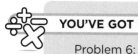
YOU'VE GOT PROBLEMS

Problem 6: Factor the expression $7x^2y - 21xy^3$.

Special Factoring Patterns

You should feel comfortable factoring trinomials such as $x^2 + 5x + 4$ using whatever method suits you. Most people play with binomial pairs until they stumble across something that works, in this case $(x + 4)(x + 1)$, whereas others undertake more complicated means. Regardless of your personal "flair," there are some patterns you should have memorized:

- Difference of perfect squares: $a^2 - b^2 = (a + b)(a - b)$

Explanation: A perfect square is a number like 16, which can be created by multiplying something times itself. In the case of 16, that something is 4, since 4 times itself is 16. If you see one perfect square being subtracted from another, you can automatically factor it using the pattern above. For example, $x^2 - 25$ is a difference of x^2 and 25, and both are perfect squares. Thus, it can be factored as $(x + 5)(x - 5)$.

KELLEY'S CAUTIONS

You cannot factor the *sum* of perfect squares, so whereas $x^2 - 4$ is factorable, $x^2 + 4$ is not!

- Sum of perfect cubes: $a^3 + b^3 = (a + b)(a^2 - ab + b^2)$

Explanation: Perfect cubes are similar to perfect squares. The number 125 is a perfect cube because $5 \cdot 5 \cdot 5 = 125$. This pattern is a little clumsier to memorize, but it can come in handy occasionally. This formula can be altered just slightly to factor the *difference* of perfect cubes, as illustrated in the next bullet. Other than a couple of sign changes, the process is the same.

- Difference of perfect cubes: $a^3 - b^3 = (a - b)(a^2 + ab + b^2)$

Explanation: Enough with the symbols for these formulas—let's do an example.

Example 6: Factor $x^3 - 27$ using the difference of perfect cubes factoring pattern.

Solution: Note that x is a perfect cube since $x \cdot x \cdot x = x^3$, and 27 is also, since $3 \cdot 3 \cdot 3 = 27$. Therefore, $x^3 - 27$ corresponds to $a^3 - b^3$ in the formula, making $a = x$ and $b = 3$. Now, all that's left to do is plug a and b into the formula:

$$a^3 - b^3 = \left(a - b\right)\left(a^2 + ab + b^2\right)$$
$$x^3 - 27 = \left(x - 3\right)\left(x^2 + 3x + 9\right)$$

You cannot factor $(x^2 + 3x + 9)$ any further, so you are finished.

YOU'VE GOT PROBLEMS

Problem 7: Factor the expression $8x^3 + 343$.

Solving Quadratic Equations

Before you put algebra in the rearview mirror, there's one last stop. Sure, you've been able to solve equations like $x + 9 = 12$ forever, but when the equations get a little trickier, maybe you get a little panicky. Forgetting how to solve quadratic equations (equations whose highest exponent is a 2) has distinct symptoms: dizziness, shortness of breath, nausea, and loss of appetite. To fight this ailment, take the following 3 tablespoons of quadratic problem solving and call me in the morning.

Every quadratic equation can be solved with the *quadratic formula* (method three, which follows), but it's important that you know the other two methods as well. Factoring is undoubtedly the fastest of the three methods, so you should try it first. Few people choose completing the square as their first option, but it (like the quadratic formula) works every time, though it requires a few more steps than its counterpart. However, you *have* to learn completing the square, because it pops up later in calculus, when you least expect it.

Method One: Factoring

To begin, set your quadratic equation equal to 0; this means add and subtract the terms as necessary to get them all to one side of the equation. If the resulting equation is factorable, factor it and set each individual term equal to 0. These little baby equations will give you the solutions to the equation. That's all there is to it.

Example 7: Solve the equation $3x^2 + 4x = -1$ by factoring.

Solution: Always start the factoring method by setting the equation equal to 0. In this case, start by adding 1 to each side of the equation: $3x^2 + 4x + 1 = 0$.

Now, factor the equation and set each factor equal to 0. This creates two cute little mini-equations that need to be solved, giving you the final answer:

$$(3x+1)(x+1)=0$$

$$3x+1=0 \qquad x+1=0$$
$$\text{or}$$
$$x=-\tfrac{1}{3} \qquad x=-1$$

This equation has two solutions: $x = -\tfrac{1}{3}$ or $x = -1$. You can check them by plugging each separately into the original equation, and you'll find that the result is true.

Method Two: Completing the Square

As I mentioned earlier, this method is a little trickier than the other two, but you really do need to learn it now, or you'll be coming back to figure it out later. I've discovered that it's best to learn this method in the context of an example, so let's go to it.

Example 8: Solve the equation $2x^2 + 12x - 18 = 0$ by completing the square.

Solution: In this method, unlike factoring, you want the constant separate from the variable terms, so move the constant to the right side of the equation by adding 18 to both sides:

$$2x^2 + 12x = 18$$

This is important: For completing the square to work, the coefficient of x^2 *must* be 1. In this case, it is 2, so to eliminate that pesky coefficient, divide every term in the equation by 2:

$$x^2 + 6x = 9$$

 KELLEY'S CAUTIONS

If you don't make the coefficient of the x^2 term 1, then the rest of the completing-the-square process will not work. Also, when you divide to eliminate the x^2 coefficient, make sure you divide *every term* in the equation (including the constant, sitting dejectedly on the other side of the equation).

Here's the key to completing the square: take half of the coefficient of the x term, square it, and add it to both sides. In this problem, the x coefficient is 6, so take half of it (3) and square that ($3^2 = 9$). Add the result (9) to both sides of the equation:

$$x^2 + 6x + 9 = 9 + 9$$
$$x^2 + 6x + 9 = 18$$

At this point, if you've done everything correctly, the left side of the equation will be factorable. In fact, it will be a perfect square!

$$(x+3)(x+3) = 18$$
$$(x+3)^2 = 18$$

To solve the equation, take the square root of both sides. That will cancel out the exponent. Whenever you do this, you have to add a ± sign in front of the right side of the equation. This is always done when square rooting both sides of any equation:

$$\sqrt{(x+3)^2} = \pm 18$$
$$x + 3 = \pm\sqrt{18}$$

To solve for x, subtract 3 from each side, and that's it. It would also be good form to simplify $\sqrt{18}$ into $3\sqrt{2}$:

$$x = -3 \pm 3\sqrt{2}$$

Method Three: The Quadratic Formula

The quadratic formula is one-stop shopping for all your quadratic equation needs. All you have to do is make sure your equation is set equal to 0, and you're halfway there. Your equation will then look like this: $ax^2 + bx + c = 0$, where a, b, and c are the coefficients as indicated. Take those numbers and plug them straight into this formula (which you should definitely memorize):

$$x = \frac{-b \pm \sqrt{b^2 - 4ac}}{2a}$$

You'll get the same answer you would achieve by completing the square. Just to convince you that the answer's the same, we'll do the problem in Example 8 again, but this time with the quadratic formula.

Example 9: Solve the equation $2x^2 + 12x - 18 = 0$, this time using the quadratic formula.

Solution: Because the equation is already set equal to 0, it is in form $ax^2 + bx + c = 0$, and $a = 2$, $b = 12$, and $c = -18$. Plug these values into the quadratic formula and simplify:

$$x = \frac{-12 \pm \sqrt{12^2 - 4(2)(-18)}}{2(2)}$$

$$x = \frac{-12 \pm \sqrt{144 - (-144)}}{4}$$

$$x = \frac{-12 \pm \sqrt{288}}{4}$$

$$x = \frac{-12 \pm 12\sqrt{2}}{4}$$

$$x = \frac{-12}{4} \pm \frac{12\sqrt{2}}{4}$$

$$x = -3 \pm 3\sqrt{2}$$

So although there are fewer steps to the quadratic formula, there is some room for error during computation. You should practice both methods, but primarily use the one that feels more comfortable to you.

YOU'VE GOT PROBLEMS

Problem 8: Solve the equation $3x^2 + 12x = 0$ three times, using three different methods: greatest common factor, completing the square, and the quadratic formula.

Synthesizing the Quadratic Solution Methods

Whenever you are learning (or reviewing) mathematical techniques, it's easy to get lost in the details. While it's true that many solution methods require you to understand and follow a series of steps, math is more than a process to follow. It is not merely a numbered list of commands to execute robotically.

You might be asking yourself, "When do I know which technique to use?" Well, you should always try factoring first, because it's usually the easiest method. If factoring doesn't work, then which should you choose: the quadratic formula or completing the square?

Honestly, the choice is yours. Select the method that feels most comfortable to you, because although the methods are very different in process, they are actually more related than they may at first appear. In fact, did you know that if you ever forget the quadratic formula, you can generate it from scratch? Just complete the square on the generic quadratic equation $ax^2 + bx + c = 0$.

Example 10: Generate the quadratic formula by completing the square for $ax^2 + bx + c = 0$.

Solution: The coefficient of x^2 must be 1 in order to complete the square, so divide each term by the current coefficient (a) and then move the constant (the term with no x-part) to the right side of the equation by subtracting:

$$\frac{ax^2}{a} + \frac{bx}{a} + \frac{c}{a} = \frac{0}{a}$$

$$x^2 + \frac{b}{a}x + \frac{c}{a} = 0$$

$$x^2 + \frac{b}{a}x = -\frac{c}{a}$$

According to the technique described in Example 8, you must take half of the x-coefficient $\left(\frac{1}{2} \cdot \frac{b}{a} = \frac{b}{2a}\right)$, square it $\left(\frac{b}{2a} \cdot \frac{b}{2a} = \frac{b^2}{4a^2}\right)$, and add the result $\left(\frac{b^2}{4a^2}\right)$ to both sides of the equation.

$$x^2 + \frac{b}{a}x + \frac{b^2}{4a^2} = -\frac{c}{a} + \frac{b^2}{4a^2}$$

The left side of the equation is a perfect square. Notice that the term inside the squared quantity is $\frac{b}{2a}$, the value that you squared just a moment ago.

CRITICAL POINT

Whenever you complete the square, the constant inside the squared quantity is *always* half of the original x-coefficient.

$$\left(x + \frac{b}{2a}\right)^2 = -\frac{c}{a} + \frac{b^2}{4a^2}$$

In order to add the fractions on the right side, you will need common denominators.

$$\left(x + \frac{b}{2a}\right)^2 = -\frac{c}{a} \cdot \frac{4a}{4a} + \frac{b^2}{4a^2}$$

$$\left(x + \frac{b}{2a}\right)^2 = -\frac{4ac}{4a^2} + \frac{b^2}{4a^2}$$

$$\left(x + \frac{b}{2a}\right)^2 = \frac{b^2 - 4ac}{4a^2}$$

You may have noticed the familiar "$b^2 - 4ac$" from the quadratic formula. You're almost done! To solve for x, take the square root of both sides of the equation.

$$\sqrt{\left(x + \frac{b}{2a}\right)^2} = \pm\sqrt{\frac{b^2 - 4ac}{4a^2}}$$

$$x + \frac{b}{2a} = \pm\frac{\sqrt{b^2 - 4ac}}{\sqrt{4a^2}}$$

$$x + \frac{b}{2a} = \pm\frac{\sqrt{b^2 - 4ac}}{\sqrt{4} \cdot \sqrt{a^2}}$$

$$x + \frac{b}{2a} = \pm\frac{\sqrt{b^2 - 4ac}}{2a}$$

Subtract $\frac{b}{2a}$ from both sides to solve for x, and you've generated the quadratic formula as if by magic!

$$x = -\frac{b}{2a} \pm \frac{\sqrt{b^2 - 4ac}}{2a}$$

$$x = \frac{-b \pm \sqrt{b^2 - 4ac}}{2a}$$

The moral of this story: although the solution methods for quadratic equations may look quite different from one another, they have a lot in common.

Before wrapping up our discussion on quadratic equations, let's review the relationship between the factors and x-intercepts of a quadratic equation.

Example 11: Create a quadratic equation that has x-intercepts $x = -2$ and $x = 5$.

Solution: First, a warning: there are many possible correct answers, but this is the easiest solution. You may want to look back at Example 7 for a moment, because you're going to follow that process in reverse.

If $x = -2$ is a solution for the quadratic equation, then you could add 2 to both sides of the equation to get an equivalent equation:

$$x = -2$$
$$x + 2 = 0$$

Similarly, you could subtract 5 from both sides of the equation $x = 5$:

$$x = 5$$
$$x - 5 = 0$$

Notice that $(x + 2)$ and $(x - 5)$ could be factors of the quadratic equation you're looking for, because they both equal zero. If one of the factors equals 0, then multiplying anything by that factor also gives you zero.

Why is this good news? If the entire equation equals 0, then you've found a root, an x-intercept. Therefore, the simplest quadratic equation with x-intercepts -2 and 5 would be the product of $(x + 2)$ and $(x - 5)$:

$$(x+2)(x-5) = x^2 - 5x + 2x - 10$$
$$= x^2 - 3x - 10$$

Quadratic equation $y = x^2 - 3x - 10$ has x-intercepts $x = -2$ and $x = 5$. If you're skeptical, substitute them back into the equation to verify. You'll get $y = 0$, which means points $(-2,0)$ and $(5,0)$ lie on the graph (and also on the x-axis).

YOU'VE GOT PROBLEMS

Problem 9: Create a quadratic equation that has x-intercepts $x = -1$ and $x = \frac{2}{3}$.

The Least You Need to Know

- Basic equation solving is an important skill in calculus.
- Reviewing the five exponential rules will prevent arithmetic mistakes in the long run.
- You can create the equation of a line with just a little information using point-slope form.
- There are three major ways to solve quadratic equations, each important for different reasons.

Equations, Relations, and Functions

I still remember the fateful day in Algebra I when the equation $y = 3x + 2$ became $f(x) = 3x + 2$. The dreaded function! At the time, I didn't quite understand why we had to make the switch. I was a fan of the y and was sad to see it go. What I failed to grasp was that the advent of the function marked a new step forward in my math career.

If you know that an equation is also a function, it guarantees that the equation in question will always behave in a certain way. Most of the definitions in calculus require functions in order to operate correctly. Therefore, the vast majority of our work in calculus will be with functions exclusively, with only a few minor exceptions. So it's good to know exactly what a function is, to be able to recognize important functions at a glance, and to be able to perform basic function operations.

In This Chapter

- When is an equation a function?
- Important function properties
- Building your function skills repertoire
- The basics of parametric equations

What Makes a Function Tick?

Let's get a little vocabulary straight before we get too far. Any sort of equation in mathematics is classified as a *relation,* as the equation describes a specific way that the variables and numbers in the equation are related. Relations don't have to be equations, although that is how they are most commonly written.

📖 DEFINITION

A **relation** is a collection of related numbers, usually described by an equation, graph, or list of ordered pairs. A **function** is a relation such that every input has only one matching output.

Here's the most basic definition of a relation. You'll notice that there's not a whole lot to it, just a list of ordered pairs:

$$s : \{(-1,5),(1,6),(2,4)\}$$

This relation, called s, gives a list of inputs and outputs. In essence, you're asking s, "What will you give me if I give you -1?" The reply is 5, because the ordered pair $(-1,5)$ appears in the relation. If you input 2, s spits back 4. However, if you input 6, s has no response; the only inputs s accepts are -1, 1, and 2, and the only outputs it can offer are 5, 6, and 4.

In calculus, it is more useful to write relations like this:

$$g(x) = \tfrac{1}{3}x - 3$$

This relation, called g, accepts any real number input. To find out the output g gives, you plug the input into the x slot. For example, if I input $x = 21$, the output—called $g(21)$—is found as follows:

$$g(21) = \tfrac{1}{3}(21) - 3$$
$$= \tfrac{21}{3} - 3$$
$$= 7 - 3$$
$$= 4$$

A *function* is a specific kind of relation. In a function, no input is allowed to give you more than one output. When one number goes in, only one matching number is allowed to come out. The relation g here is a function of x, because for every x you plug in, you can only get one result. If you plug in $x = 3$, you will always get -2. If you did it 50 times, you wouldn't suddenly get 101.7 as your answer on the forty-ninth try! Every input results in only one corresponding output. Different inputs can result in different outputs, for example, $g(3) \neq g(6)$. That's okay. You just can't get different answers when you plug in the same initial quantity.

✏️ CRITICAL POINT

A function does not *have* to have a name, like $f(x)$ or $h(x)$, to be a function. The relations $y = x^2$ and $f(x) = x^2$ are equally qualified to be functions even though they look different.

The word *domain* is usually used to describe the set of inputs for a function. Any number that a function accepts as an appropriate input is part of the domain. For example, in the function s:{(–1,5),(1,6),(2,4)}, the domain is {–1,1,2}. The set of outputs to a function is called the *range*. The range of s is {4,5,6}.

Enough math for a second—let's relate this to real life. A person's height is a function of time. If I ask, "How tall were you at exactly noon today?" you could give only one answer. You couldn't respond "5 feet 6 inches" *and* "6 feet 1 inch," unless, of course, you lied on your driver's license.

Sometimes you'll plug more than a number into a function—you can also plug a *function* into another function. This is called composition of functions, and is not difficult to do. Simply start by evaluating the inner function and work your way out.

Example 1: If $f(x) = \sqrt{x}$ and $g(x) = x + 6$, evaluate $g(f(25))$.

Solution: In this case, 25 is plugged into f, and that output is in turn plugged into g. Start in the belly of the beast and evaluate $f(25)$. This is easy: $f(25) = \sqrt{25} = 5$. Now, plug this result into g:

$$g(5) = 5 + 6 = 11$$

Therefore, $g(f(25)) = 11$.

YOU'VE GOT PROBLEMS

Problem 1: If $f(x) = \frac{x-1}{6}$, $g(x) = x^2 + 15$, and $h(x) = \sqrt[3]{x}$, evaluate $h(g(f(43)))$.

Sometimes in calculus, you run across a weird entity: the piecewise-defined function. This function is similar to Frankenstein's monster because it is created by sewing other functions together. The next example explains how to interpret and evaluate piecewise-defined functions.

Example 2: Given the piecewise-defined function $f(x)$ defined below, calculate $f(–1)$, $f(2)$, and $f(10)$. Then, draw the graph of $f(x)$.

$$f(x) = \begin{cases} 2x+3, & x < 2 \\ x-4, & x \geq 2 \end{cases}$$

Solution: A piecewise-defined function like $f(x)$ uses more than one expression to generate its values. In this case, you will either substitute values of x into $2x + 3$ or $x – 4$. How do you know which one to use? It depends on the number, x, you're substituting in.

Notice the inequality statements attached to the expressions, such as $x < 2$ next to the expression $2x + 3$. Use this expression for any x-value less than 2. For example, $x = -1$ qualifies:

$$f(-1) = 2(-1) + 3$$
$$= -2 + 3$$
$$= 1$$

Use the other expression in the piecewise-defined function, $x - 4$, for all x-values greater than or equal to 2. Both $f(2)$ and $f(10)$ represent such x-values.

$$f(2) = 2 - 4 \qquad f(10) = 10 - 4$$
$$= -2 \qquad\qquad = 6$$

The graph of $f(x)$ consists of two pieces, one for each expression. Begin by graphing the line $y = 2x + 3$, which has y-intercept 3 and slope $m = 2$. However, the graph only applies when $x < 2$, so erase any part of the graph to the right of $x = 2$.

Next, graph $y = x - 4$ (a line with y-intercept -4 and slope $m = 1$). This portion of the graph only applies when $x \geq 2$, so erase the portion of the graph left of $x = 2$. The finished graph appears in Figure 3.1.

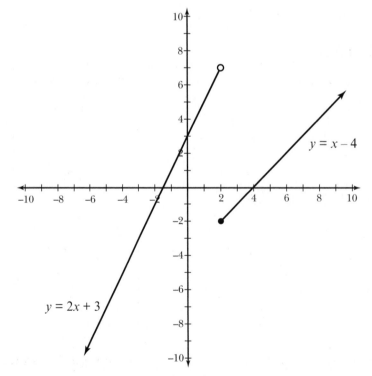

Figure 3.1
The graph of piecewise-defined function f(x). *Note the open and closed dots where the graph splits at* x = 2.

Notice that $y = 2x + 3$ ends in an open dot at the point (2,7). The graph does not actually include this point, because $2x + 3$ only applies when x is less than 2, *not* when x equals 2. However, the graph of $y = x - 4$ starts in a closed dot, because that graph applies to x-values greater than *or equal to* 2. An open dot indicates an excluded point, and a closed dot indicates an included point.

YOU'VE GOT PROBLEMS

Problem 2: Given the piecewise-defined function $g(x)$ defined here, calculate $g(-2)$, $g(0)$, and $g(5)$:

$$g(x) = \begin{cases} 12 - x^2, & x \leq 0 \\ \sqrt{x^2 - 9} & x > 0 \end{cases}$$

The last important thing you should know about functions is the *vertical line test*. This test is a way to tell whether a given graph is the graph of a function or not. All you have to do is draw imaginary vertical lines through the graph and note the number of times these lines hit the graph (see Figure 3.2). If any imaginary line can be drawn through the graph that hits it more than once, the graph cannot be a function.

DEFINITION

The **vertical line test** tells you whether or not a graph is a function. If any vertical line can be drawn through the graph that intersects that graph more than once, then the graph in question cannot be a function.

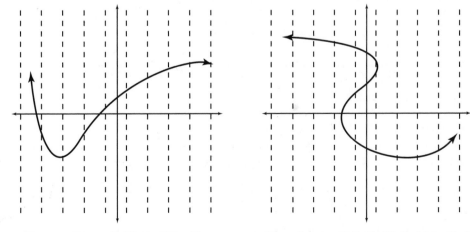

Function: Passes the Vertical Line Test   *Not a function:* Fails the Vertical Line Test

Figure 3.2
No vertical line intersects the graph on the left more than once, so it is a function. However, some vertical lines hit the right-hand graph more than once, so it cannot be a function.

Working with Graphs of Functions

You don't need to know the expressions that define functions in order to perform basic operations on them. You can conduct a lot of work given graphs or even tables of values. All of the same basic rules apply.

Example 3: Given the graph of $f(x)$ in Figure 3.3, identify the domain and range of $f(x)$ and calculate $f(-4)$.

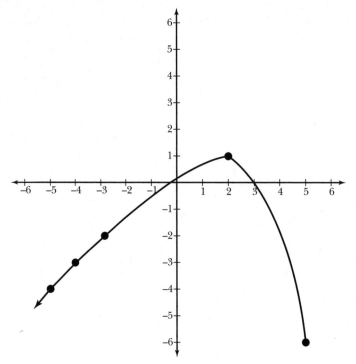

Figure 3.3

The graph of a function f(x) *and an assortment of points on the graph. Assume that the coordinates of the specified points have integer (nonfraction) values.*

Solution: To determine the domain of a function based on its graph, imagine a vertical line sweeping along the graph from left to right. As that line intersects the graph, the intersections represent possible x-values for the function, so they belong in the domain.

For example, sketch the vertical line $x = -5$ on Figure 3.3. It intersects $f(x)$ at the point $(-5,-4)$. Therefore, $x = -5$ belongs in the domain of $f(x)$.

In fact, any vertical line will intersect the graph of $f(x)$ until you get to $x = 5$. The graph abruptly ends at point $(5,-6)$; there is no arrow there indicating that the graph will continue infinitely downward. Therefore, the domain of $f(x)$ consists of all real numbers less than or equal to 5.

To determine the range, use a similar method: imagine a horizontal line sweeping from the bottom to the top of the graph. Any time the line intersects the graph, that y-value belongs in the range. Because the graph only reaches a height of $y = 1$, the range of the graph is all real numbers less than or equal to 1.

Finally, to calculate $f(-4)$, find the point on the graph whose x-coordinate is -4. Notice that the point $(-4,-3)$ lies on the graph. Therefore, according to the graph, $f(-4) = -3$.

If you got an answer of -5, don't panic. You just mixed up the values of x and y. The graph does pass through the point $(-5,-4)$, but that means $f(-5) = -4$. In that case, -4 is the *output* when -5 is the input. The problem asks you to calculate $f(-4)$; it wants to know what the output is when -4 is the input.

Example 4: Consider the functions $g(x)$ and $h(x)$ presented in Figure 3.4. You are given the graph of $g(x)$ and a table that includes some of the function values of $h(x)$. Based on the information given, calculate $g(h(6))$ and $h(g(4))$.

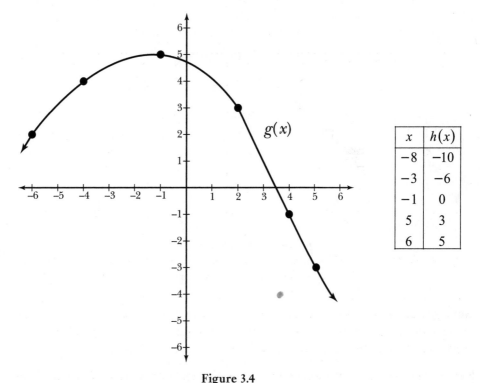

x	$h(x)$
-8	-10
-3	-6
-1	0
5	3
6	5

Figure 3.4

Two functions are represented here, g(x) as a graph and h(x) as a table. Note that the specified points on g(x) have integer coordinates.

Solution: Don't be intimidated by the strange way these functions are presented. Whether they're expressions, graphs, or charts, functions are just a relationship between pairs of numbers, between inputs and outputs.

In this example, the graph of $g(x)$ highlights six of its function values:

$$g(-6)=2 \qquad g(-1)=5 \qquad g(4)=-1$$
$$g(-4)=4 \qquad g(2)=3 \qquad g(5)=-3$$

The chart in Figure 3.4 reports five function values of $h(x)$, in a very straightforward manner.

$$h(-8)=-10 \qquad h(-1)=0 \qquad h(6)=5$$
$$h(-3)=-6 \qquad h(5)=3$$

This is all the information you need to solve both problems. It's time to calculate $g(h(6))$. First things first, work from the inside out, determining the value of $h(6)$. According to the value list you just made, $h(6) = 5$. Replace $h(6)$ with the equivalent expression, 5.

$$g(h(6)) = g(5)$$

According to the list of function values for $g(x)$, $g(5) = -3$. Therefore, $g(h(6)) = -3$.

Think you've got it? Try the next problem on for size: $h(g(4))$. According to the list of values for $g(x)$, $g(4) = -1$. That means $h(g(4)) = h(-1)$. Note that $h(-1) = 0$. Therefore, you conclude that $h(g(4)) = 0$.

YOU'VE GOT PROBLEMS

Problem 3: Given $p(x) = 4 - x$ and the graph of $q(x)$ in Figure 3.5, calculate $p(q(1))$.

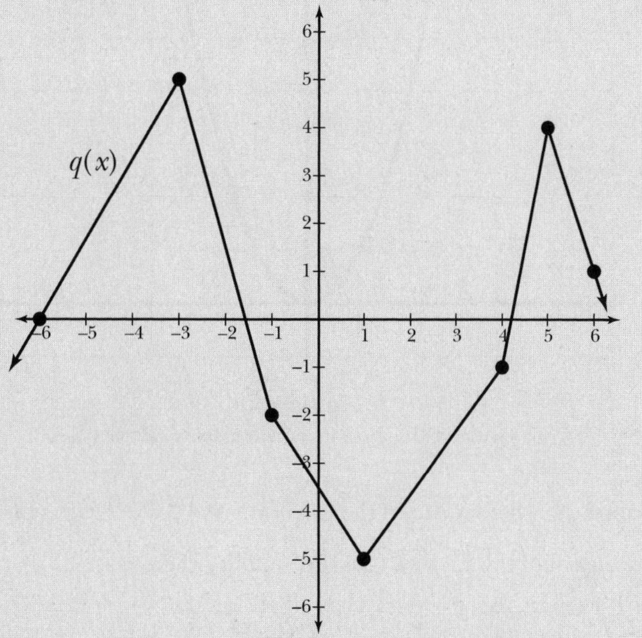

Figure 3.5
Like the preceding examples in this chapter, all highlighted points on the graph of q(x) *have integer coordinates.*

Functional Symmetry

Now that you know a thing or two about functions, you should also know some of the key classifications and buzzwords. If you throw these words around at parties, you'll surely wow your friends. Just think about how impressed they'd be with an offhand comment like, "That painting really exploits y-symmetry to show us our miniscule place in the world." Maybe you and I don't go to the same sorts of parties ….

A function is *symmetric* if it mirrors itself with respect to a fixed part of the coordinate plane. That sounds like a complicated concept, but it isn't. Consider, for example, the graph of $y = x^2$.

DEFINITION

A **symmetric** function looks like a mirror image of itself, typically across the x-axis, y-axis, or about the origin.

In Figure 3.6, notice that the graph looks exactly the same on either side of the *y*-axis. This function is said to be *y*-symmetric. There is an easy arithmetic test for *y*-symmetry that doesn't require the graph.

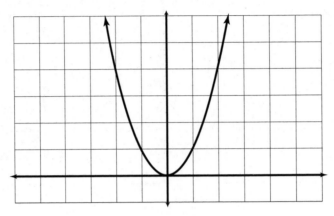

Figure 3.6
Feast your eyes on a graph that is symmetric about the y-*axis.*

Example 5: Determine whether or not the graph $y = x^4 - 2x^2 + 1$ is *y*-symmetric.

Solution: Replace each of the *x*'s with (–*x*) and simplify the equation:

$$y = \left(-x\right)^4 - 2\left(-x\right)^2 + 1$$
$$y = x^4 - 2x^2 + 1$$

Whenever a negative number is raised to an even power, the negative sign will be eliminated. Notice that our simplified result is the same as the original equation. When this happens, you know that the equation is, indeed, *y*-symmetric. (By the way, *y*-symmetric functions are also classified as *even* functions.) In case you'd also like visual proof, check out the graph in Figure 3.7:

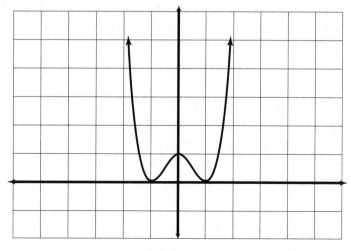

Figure 3.7
The graph of $y = x^4 - 2x^2 + 1$.

The other two major kinds of symmetry are *x*-symmetry and origin-symmetry, illustrated by the graphs in Figure 3.8.

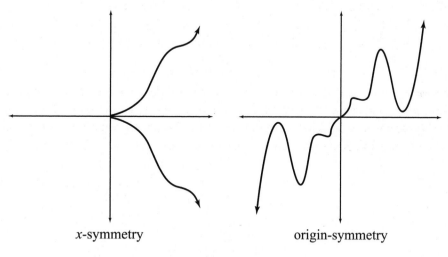

x-symmetry   origin-symmetry

Figure 3.8
Two other types of symmetry you may encounter. Note that most x-*symmetric equations are not functions, because they fail the vertical line test.*

Very similar to *y*-symmetry, *x*-symmetry requires that the graph be identical above and below the *x*-axis. The test for *x*-symmetry is also similar to *y*-symmetry, except that you plug in (–*y*) for the *y*'s instead of (–*x*) for the *x*'s. Again, if the equation reverts to its original form when simplifying is over, then the equation is *x*-symmetric. If even one sign is different, the equation is not *x*-symmetric.

Origin-symmetry is achieved when the graph does exactly the opposite thing on either side of the origin. In Figure 3.8, notice that the origin-symmetric curve snakes up and to the right as *x* gets more positive, and it heads down and to the left as *x* gets more negative. In fact, every turn in the first quadrant is matched and inverted in the third quadrant.

To test an equation for origin-symmetry, replace all *x*'s with (–*x*) and all *y*'s with (–*y*). Once again, if the simplified equation matches your original equation, then that function is origin-symmetric. By the way, if a function is origin-symmetric, you can also classify it as an odd function.

Example 6: Demonstrate algebraically that the function $y = 2x^3 - x$ is origin-symmetric.

Solution: Replace *y* with –*y* and replace each *x* with –*x*:

$$-y = 2(-x)^3 - (-x)$$

Simplify the equation.

$$-y = -2x^3 + x$$

If the function is truly origin-symmetric, then solving it for *y* (rather than –*y* as it currently appears) will produce the original equation. Multiply all of the terms by –1 to solve for *y*.

$$-1(-y) = (-1)(-2x^3) + (-1)(x)$$
$$y = 2x^3 - x$$

The result matches the original equation, so $y = 2x^3 - x$ is origin-symmetric. For visual proof, check out Figure 3.9.

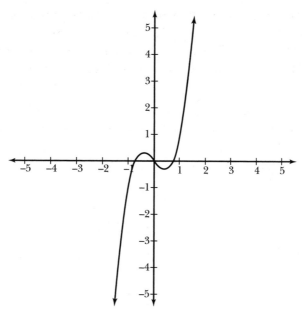

Figure 3.9
The origin-symmetric graph of y = 2x³ − x.

YOU'VE GOT PROBLEMS

Problem 4: Determine what kind of symmetry, if any, is evident in the graph of $y = \frac{x^3}{|x|}$.

Graphs to Know by Heart

During your study of calculus, you'll see certain graphs over and over again. Because of this, it's important to know them intuitively. You're already familiar with these functions, but make sure you know their graphs intimately, and it will save you time and frustration in the long run. Tell that someone special in your life that they can no longer possess your whole heart—they're going to have to share it with some math graphs. If they don't understand your needs, it wasn't meant to be for the two of you.

The descriptions are as follows:

- $y = x$: the most basic linear equation; has slope 1 and y-intercept 0; origin-symmetric; both domain and range are all real numbers

- $y = x^2$: the most basic quadratic equation; y-symmetric; domain is all real numbers; range is $y \geq 0$

- $y = x^3$: the most basic cubic equation; origin-symmetric; domain and range are all real numbers

- $y = |x|$: the absolute value function; returns the positive form of the input; y-symmetric; made of two line segments of slope −1 and 1, respectively; domain is all real numbers; range is $y \geq 0$

- $y = \sqrt{x}$: the square root function; has no symmetry; domain is $x \geq 0$ (you can't find the square root of numbers less than 0); range is $y \geq 0$

- $y = \frac{1}{x}$: no x- or y-intercepts; origin-symmetric; domain and range are both all real numbers except for 0

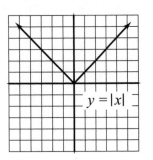
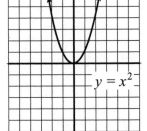
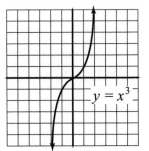
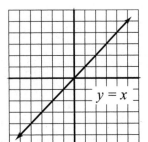
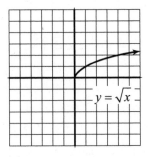
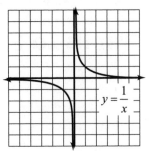

Figure 3.10

The six most basic functions that will soon reside in your heart (specifically the left ventricle).

Constructing an Inverse Function

You've used inverse functions forever without even realizing it. They are the tools you break out to eliminate something unwanted in an equation. For example, how would you solve the equation $x^2 = 9$? To solve for x, you would take the square root of both sides to eliminate the squared term. This works because $y = \sqrt{x}$ and $y = x^2$ are inverse functions.

Mathematically speaking, f and g are inverse functions if composing the two functions in any order produces x:

$$f(g(x)) = g(f(x)) = x$$

In other words, plugging g into f and f into g leaves behind no trace of the function (not even forensic evidence), only x. Let's go back to $y = \sqrt{x}$ and $y = x^2$ for a second and show mathematically that they are inverse functions. If we plug these functions into each other, they will cancel out, leaving only x behind:

$$\left(\sqrt{x}\right)^2 = \sqrt{x^2} = x$$

YOU'VE GOT PROBLEMS

Problem 5: Verify mathematically that $y = \frac{1}{2}x^3 - 3$ and $y = \sqrt{2x + 6}$ are inverse functions using composition of functions.

Inverse functions have special notation. The inverse to a function $f(x)$ is written as $f^{-1}(x)$. This does *not* mean "f to the −1 power." It is read "the inverse of f" or "f inverse." I know the notation is a little confusing, because a negative exponent usually means that the indicated piece belongs in a different part of the fraction.

Now for some good news. It's easy to create an inverse function. The word "easy" is usually misleading when used by math teachers. In fact, whenever I qualified a class discussion with "Now, this is easy …," the students knew that it was going to be anything but. However, I wouldn't lie to you, would I? You decide as you read the next example.

Example 7: If $g(x) = \sqrt[3]{2x + 5}$, find $g^{-1}(x)$.

Solution: For starters, replace the function notation $g(x)$ with y:

$$y = \sqrt[3]{2x + 5}$$

Here's the key step: reverse the x and y. In essence, this is what an inverse function does—it turns a function inside out so that the result has the spiffy property of canceling out the initial equation:

$$x = \sqrt[3]{2y + 5}$$

Your goal now is to solve this equation for y, and you'll be done. In this problem, that means raising both sides of the equation to the third power:

$$x^3 = 2y + 5$$

Now, subtract 5 from both sides and divide by 2 to finish solving for y:

$$x^3 - 5 = 2y$$
$$\frac{x^3 - 5}{2} = y$$

That is the inverse function. To finish, write it in proper inverse function notation:

$$g^{-1}(x) = \frac{x^3 - 5}{2}$$

YOU'VE GOT PROBLEMS

Problem 6: Find the inverse function of $h(x) = \frac{2}{3}x + 5$.

In the next example, you'll work with the inverse of a function that's defined by a chart. Just to spice things up a bit, it also throws in a review of composition of functions from earlier in the chapter.

Example 8: Calculate $m^{-1}(p(3))$, assuming that functions $m(x)$ and $p(x)$ have inverse functions and that the tables below present a selection of their values.

x	$m(x)$
−2	−6
−1	−3
0	−1
2	4
3	7

x	$p(x)$
−6	5
−3	1
−1	−2
1	−4
3	−6

Solution: Remember to work from the inside out when composing two functions. In other words, begin with the function that's plugged into the other function. In this case, you begin by calculating $p(3)$. According to the $p(x)$ table, $p(3) = -6$.

Now you know that $m^{-1}(p(3)) = m^{-1}(-6)$. However, you are only given a table for $m(x)$, not for its inverse. Remember that an inverse function simply reverses the x- and y-coordinates of a function—the inputs become outputs and vice versa. Therefore, by reversing the columns of the $m(x)$ table, you can create a table of values for its inverse function.

x	$m^{-1}(x)$
−6	−2
−3	−1
−1	0
4	2
7	3

According to this new table, $m^{-1}(-6) = -2$. Therefore, you conclude that $m^{-1}(p(3)) = -2$.

Parametric Equations

With all this talk about functions, you might be leery of nonfunctions. Don't get all closed-minded on me. You can use something called *parametric equations* to express graphs, too, and they have the unique ability to represent nonfunctions (like circles) very easily. Parametric equations are pairs of equations, usually in the form of "$x =$" and "$y =$," that define points of the graph in terms of yet another variable, usually t.

DEFINITION

Parametric equations define a graph in terms of a third variable, or parameter.

What's a Parameter?

That definition's quite a mouthful, I know. To get a better understanding, let's look at an example of parametric equations:

$$x = t + 1$$
$$y = t - 2$$

These two equations together produce one graph. To find that graph, you have to substitute a spectrum of things for the parameter t; each time you make a t substitution, you'll get a point on the graph. So a parameter is just a variable into which you plug numeric values to find coordinates on a parametric equation graph. For example, if you plug $t = 1$ into the equations, you get the following:

$$x = t + 1 = 1 + 1 = 2$$
$$y = t - 2 = 1 - 2 = -1$$

Therefore, the point (2,–1) is on the graph. To get another point, I'll plug in $t = -2$, but you can actually plug in any real number for t:

$$x = -2 + 1 = -1$$
$$y = -2 - 2 = -4$$

A second point on the graph is (–1,–4). You can see that this process takes a while. In fact, it seems like only an infinite number of t-values will get you the exact graph.

Converting to Rectangular Form

Let's be honest, no one wants to plug in an infinite number of points. Even if you had the time to do that, you could definitely find something better to do. Therefore, it behooves us to learn how to translate from parametric form to the form we know and love, rectangular form. In the next example, we'll translate that set of parametric equations into something more manageable.

Example 9: Translate the parametric equations $x = t + 1$, $y = t - 2$ into rectangular form.

Solution: Begin by solving one of the equations for t. They're both pretty basic, so it doesn't matter which you choose. I'll pick the x equation so my result is in the form "$y =$." That makes it easier to graph:

$$x = t + 1$$
$$x - 1 = t$$

Now you have t in terms of x. Therefore, you can replace the t in the y equation with $(x - 1)$, because you know that $t = x - 1$:

$$y = t - 2$$
$$y = (x - 1) - 2$$
$$y = x - 3$$

This is just a line in slope-intercept form, so your parametric equations' graph is the line with slope 1 and y-intercept –3. It's graphed in Figure 3.11.

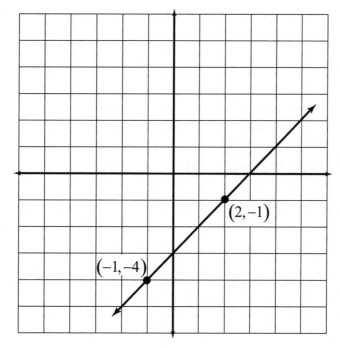

Figure 3.11

The graph of y = x − 3. *Note that the points (2,−1) and (−1,−4) are both on the graph,*
as we suspected from our work preceding Example 9.

YOU'VE GOT PROBLEMS

Problem 7: Put the parametric equations $x = t + 1$, $y = t^2 - t + 1$ into rectangular form.

The Least You Need to Know

- A relation becomes a function when each of its inputs can only result in one matching output.
- The inputs of a function comprise the domain and the outputs make the range.
- When a function is plugged into its inverse function (and vice versa), they cancel each other out.
- Parametric equations are defined by "$x =$" and "$y =$" equations that contain a parameter, usually t.

CHAPTER 4

Trigonometry: Last Stop Before Calculus

Trigonometry, the study of triangles, has been around for a long time, creeping mysteriously from shadow to shadow and occasionally snatching unwary students into its razor-sharp clutches and causing the end of their mathematics careers. Few things cause people to panic like trig does, with the exception of TV weatherman Al Roker.

It is a commonly held belief that children on All Hallow's Eve historically have marched from door to door, sacks in hand, chiming, "Trig or treat!" In response, homeowners would reward them with small protractors and compasses to avoid the wrath of neighborhood pranksters. If all the children went away happy, it was a good "sine" for harvest. However, this myth is definitely untrue, and I have gone off on a tangent.

In This Chapter

- Characteristics of periodic functions
- The six trigonometric functions
- The importance of the unit circle
- Key trigonometric formulas and identities

Getting Repetitive: Periodic Functions

There are six major trigonometric functions, at least three of which you have probably heard: sine, cosine, and tangent. All of the trigonometric functions (even those that are offended because you haven't heard of them) are periodic functions.

A *periodic function* has the unique characteristic that it repeats itself after some fixed period of time. Think of the rising of the sun as a periodic function—every 24 hours (a fixed amount of time) the sun appears on the horizon.

DEFINITION

A **periodic function**'s values repeat over and over, at the same rate and at the same intervals in time. The length of the horizontal interval after which the function repeats is called the **period.**

The amount of horizontal space it takes until the function repeats itself is called the *period*. For the most basic trigonometric functions (sine and cosine), the period is 2π. Look at the graph in Figure 4.1 of one period of $y = \sin x$.

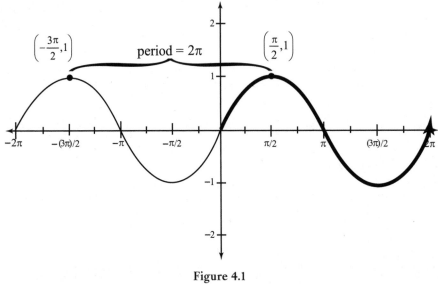

Figure 4.1
One period of y = *sin* x.

The graph of the sine function is a wave, reaching a maximum height of 1 and a minimum height of -1. On the piece of the graph shown earlier, the maximum height is reached at $x = -\frac{3\pi}{2}$ and $x = \frac{\pi}{2}$. The distance between these two points, where the graph repeats its value, is 2π. If that doesn't help you understand what is meant by period, consider the darkened portion of the graph.

This piece begins at the origin $(0,0)$ and wiggles up and down, returning to a height of 0 when $x = 2\pi$. True, the graph hits a height of 0, repeating its value, when $x = \pi$, but it hasn't completed its period yet—that is only finished at $x = 2\pi$.

If you were to extend the graph of the sine function infinitely right and left, it would redraw itself every 2π. Because of this property of periodic functions, you can list an infinite number of inputs that have identical sine values. These are called *coterminal angles,* and the next example focuses on them.

> **DEFINITION**
>
> **Coterminal angles** have the same function value, because the space between them is a multiple of the function's period.

Example 1: List two additional angles (one positive and one negative) that have the same sine value as $\frac{\pi}{4}$.

Solution: We know that sine repeats itself every 2π, so exactly 2π further up and down the x-axis from $\frac{\pi}{4}$, the value will be the same. To find these values, simply add 2π to $\frac{\pi}{4}$ in order to get one and subtract 2π from $\frac{\pi}{4}$ to get the other. In order to add and subtract the values, you'll have to get common denominators:

$$\frac{\pi}{4} + 2\pi = \frac{\pi}{4} + \frac{8\pi}{4} \qquad \frac{\pi}{4} - 2\pi = \frac{\pi}{4} - \frac{8\pi}{4}$$
$$= \frac{9\pi}{4} \qquad\qquad = -\frac{7\pi}{4}$$

Therefore, the angles $\frac{9\pi}{4}$ and $-\frac{7\pi}{4}$ are coterminal to $\frac{\pi}{4}$ and $\sin\frac{9\pi}{4} = \sin\left(-\frac{7\pi}{4}\right) = \sin\frac{\pi}{4}$.

> **CRITICAL POINT**
>
> Unless I specifically indicate otherwise, assume all angles in this book are measured in radians.

Introducing the Trigonometric Functions

Time to meet the cast. There are six players in the drama we call trigonometry. You'll see a graph of each and learn a little something about the function. Whereas it's not extremely important to memorize the graphs of these functions, it's good to see how the graphs illustrate the functions' properties.

One note before we begin. Throughout this book, I will refer to and evaluate trigonometric values in terms of radians, as they are used far more prevalently than degrees in calculus. Both degrees and radians are simply alternate ways to measure angles, just as Celsius and Fahrenheit are alternate ways to measure temperature. To get a rough idea of the conversion, remember that π radians = 180 degrees.

If you want to convert from radians to degrees, multiply by $\frac{180}{\pi}$. For example, $\frac{\pi}{2}$ is equivalent to $\frac{\pi}{2} \cdot \frac{180}{\pi} = 90$ degrees. To convert from degrees to radians, multiply by $\frac{\pi}{180}$.

Now, back to the six essential trigonometric functions, in roughly the order of importance to you in your quest for calculus.

Sine (Written as y = sin x)

The sine function is defined for all real numbers, and this unrestricted domain makes the function very trustworthy and versatile (see Figure 4.2). The range is $-1 \leq y \leq 1$, so all sine values fall within those boundaries. Notice that the sine function has a value of 0 whenever the input is a multiple of π. Sometimes, people get confused when memorizing unit circle values (more on the unit circle later in this chapter). If you remember the graph of sine, you can easily remember that $\sin 0 = \sin \pi = \sin 2\pi = 0$, because that's where the graph crosses the x-axis. The period of the sine function is 2π.

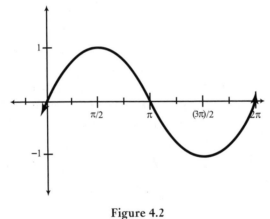

Figure 4.2
y = *sin* x.

Cosine (Written as y = cos x)

Cosine is the cofunction of sine (see Figure 4.3). (In other words, their names are the same, except one has a "co-" prefix, but I bet you figured that out.) As such, it looks very similar, possessing the same domain, range, and period. In fact, if you shift the entire graph of $y = \cos x$ a total of $\frac{\pi}{2}$ radians to the right, you get the graph of $y = \sin x$! The cosine has a value of 0 at all the "half-π's," such as $\frac{\pi}{2}$ and $\frac{3\pi}{2}$.

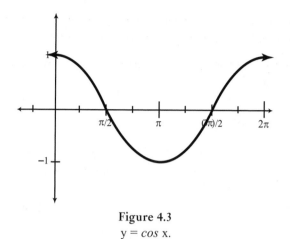

Figure 4.3
y = *cos* x.

Tangent (Written as *y* = tan *x*)

The tangent is defined as the quotient of the previous two functions: $\tan x = \frac{\sin x}{\cos x}$. Thus, to evaluate $\tan \frac{\pi}{4}$, you'd actually evaluate $\frac{\sin(\pi/4)}{\cos(\pi/4)}$ (which will equal 1 for those of you who are curious, but more about that later). Because the cosine appears in the denominator, the tangent will be undefined whenever the cosine equals 0, which (according to the last section) is at the half-π's (see Figure 4.4). Notice that the graph of the tangent has vertical *asymptotes* at these values. The tangent equals 0 at each midpoint between the asymptotes. The domain of the tangent excludes the "half-π's," $\left\{\ldots,-\frac{3\pi}{2},-\frac{\pi}{2},\frac{\pi}{2},\frac{3\pi}{2},\ldots\right\}$, but the range is all real numbers. The period of the tangent is π—notice that there's a full copy of one tangent period between $-\frac{\pi}{2}$ and $\frac{\pi}{2}$.

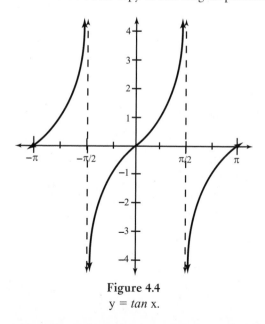

Figure 4.4
y = *tan* x.

In case you're wondering, an asymptote is a line representing an unattainable value that shapes a graph. Because the graph cannot achieve the value, the graph typically bends toward that line forever and ever, yearning, stretching, but unable to reach it. A vertical asymptote typically indicates the presence of 0 in the denominator of a fraction. For example, the vertical line $x = \frac{\pi}{2}$ is a vertical asymptote of $y = \tan x$ because the tangent has 0 in the denominator whenever $x = \frac{\pi}{2}$.

Cotangent (Written as $y = \cot x$)

The cofunction of tangent, cotangent, is the spitting image of tangent, with a few exceptions (see Figure 4.5). It, too, is defined by a quotient: $\cot x = \frac{\cos x}{\sin x}$. In fact, the cotangent is technically the *reciprocal* of the tangent, so you can also write $\cot x = \frac{1}{\tan x}$. Therefore, this function is undefined whenever $\sin x = 0$, which occurs at all the multiples of π: $\{\ldots, -2\pi, -\pi, 0, \pi, 2\pi, \ldots\}$, so the domain includes all real numbers except that set. The range, like that of the tangent, is all real numbers, and the period, π, also matches the tangent's.

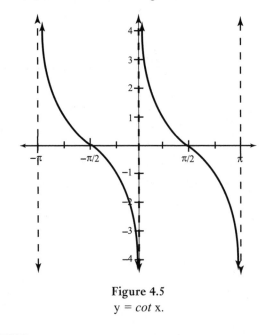

Figure 4.5
$y = \cot x.$

DEFINITION

The **reciprocal** of a fraction is the fraction flipped upside down (for example, the reciprocal of $\frac{7}{4}$ is $\frac{4}{7}$). The word *refliprocal* helps me remember what it means.

Secant (Written as $y = \sec x$)

The secant function is simply the reciprocal of cosine, so $\sec x = \frac{1}{\cos x}$. Therefore, the graph of the secant is undefined (has vertical asymptotes) at the same places (and for the same reasons) as the tangent, since they both have the same denominator (see Figure 4.6). Hence, the two functions also have the same domain. Notice that the secant has no x-intercepts. In fact, it doesn't even come close to the x-axis, only venturing as far in as 1 and −1. That's a fascinating comparison: the cosine has a range of $-1 \leq y \leq 1$, but the secant has a range of $y \leq -1$ or $y \geq 1$—almost the exact opposite. Because secant is based directly on cosine, the functions have the same period, 2π.

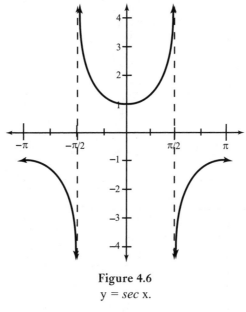

Figure 4.6
$y = sec$ x.

 **CRITICAL POINT**

It's important to know how 0 affects a fraction. If 0 appears in the denominator of a fraction, that fraction is deemed "undefined." It is against math law to divide by 0.

Cosecant (Written as $y = \csc x$)

Very similar to its cofunction sister, this function has the same range and period as the secant, differing only in its domain. Because the cosecant is defined as the reciprocal of the sine, $\csc x = \frac{1}{\sin x}$, cosecant will have the same domain as cotangent, as they share the same denominator (see Figure 4.7).

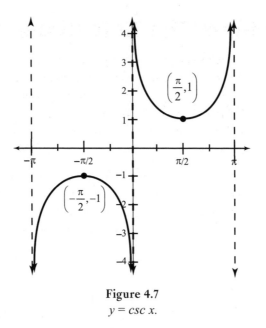

Figure 4.7
$y = \csc x.$

In essence, four of the trig functions are based on the other two (sine and cosine), so those two alone are sufficient to generate values for the rest.

KELLEY'S CAUTIONS

The cosecant is not the reciprocal of cosine. Many times, people pair these because they have the same initial *co-* sound, but that's incorrect. Similarly, the secant is not the reciprocal of sine.

Example 2: If $\cos\theta = \frac{1}{3}$ and $\sin\theta = -\frac{\sqrt{8}}{3}$, evaluate $\tan\theta$ and $\sec\theta$.

Solution: Let's tackle these one at a time. First of all, you know that $\tan\theta = \frac{\sin\theta}{\cos\theta}$, so:

$$\tan\theta = \frac{-\frac{\sqrt{8}}{3}}{\frac{1}{3}}$$

Multiply the top and bottom by 3 to simplify the fraction:

$$\tan\theta = \frac{-\frac{\sqrt{8}}{3}}{\frac{1}{3}} \cdot \frac{\frac{3}{1}}{\frac{3}{1}} = -\sqrt{8}$$

Now, on to $\sec\theta$—this is even easier. Because you know that $\cos\theta = \frac{1}{3}$, and $\sec\theta = \frac{1}{\cos\theta}$ (because the secant is the reciprocal of the cosine):

$$\sec\theta = \frac{1}{1/3} = 3$$

What's Your Sine: The Unit Circle

No one expects you to be able to evaluate most trigonometric expressions off the top of your head. If someone held a gun to my head and asked me to evaluate $\cos \frac{3\pi}{7}$ with an accuracy of .001, I would respond by calmly lying on the ground, drawing a chalk outline around myself, and preparing for death. I'd have no chance without a calculator or a Rain Man–like ability for calculation. Most calculus classes, however, will require you to know certain trigonometric values without a second thought.

These values are derived from something called the *unit circle,* a circle with a radius of length 1 that generates common cosine and sine values. You don't really have to know how to get those values (or how the unit circle works), but you should have these values memorized. Make flash cards, recite them with a partner, get a tattoo—whatever method you use to remember things— but memorize the unit circle values in the chart in Figure 4.8.

📖 **DEFINITION**

The **unit circle** is a circle whose radius is 1 unit that can be used to generate the most common values of sine and cosine. Rather than generating them each time you need them, it's best to simply memorize those common values.

angle	cosine	sine	angle	cosine	sine
0	1	0	π	-1	0
$\pi/6$	$\sqrt{3}/2$	$1/2$	$7\pi/6$	$-\sqrt{3}/2$	$-1/2$
$\pi/4$	$\sqrt{2}/2$	$\sqrt{2}/2$	$5\pi/4$	$-\sqrt{2}/2$	$-\sqrt{2}/2$
$\pi/3$	$1/2$	$\sqrt{3}/2$	$4\pi/3$	$-1/2$	$-\sqrt{3}/2$
$\pi/2$	0	1	$3\pi/2$	0	-1
$2\pi/3$	$-1/2$	$\sqrt{3}/2$	$5\pi/3$	$1/2$	$-\sqrt{3}/2$
$3\pi/4$	$-\sqrt{2}/2$	$\sqrt{2}/2$	$7\pi/4$	$\sqrt{2}/2$	$-\sqrt{2}/2$
$5\pi/6$	$-\sqrt{3}/2$	$1/2$	$11\pi/6$	$\sqrt{3}/2$	$-1/2$

Figure 4.8
Values for the confounded unit circle, a necessary evil to calculus. Memorize it now, and avoid trauma in the future. All angles are measured in radians.

If you're having trouble remembering the unit circle, look for patterns. If you absolutely refuse to memorize these values and it's okay with your instructor that you don't, at the very least keep this chart in a handy place, because you'll find yourself consulting it often.

Now that you know the unit circle and all kinds of crazy stuff about trig functions, your powers have increased. (Just make sure you always use them for good, not evil.) In fact, you are able to evaluate a lot more functions, as demonstrated by the next example.

Example 3: Find the value of $\cos \frac{23\pi}{4}$ without using a calculator.

Solution: You only know the values of sine and cosine from 0 radians to 2π radians.

Clearly, $\frac{23\pi}{4}$ is much too large to fit in this limited interval. However, because cosine is a periodic function, its values will repeat. Since cosine's period is 2π, you can find a coterminal angle to $\frac{23\pi}{4}$, which does appear in our unit circle chart, and evaluate that one instead—the answer will be the same.

According to Example 1 in this chapter, all you have to do is add or subtract the period (again, it is 2π for cosine) and you'll get a coterminal angle. I'm looking for a smaller angle than $\frac{23\pi}{4}$, so I'll subtract 2π. Don't forget to get common denominators to subtract correctly:

$$\frac{23\pi}{4} - 2\pi = \frac{23\pi}{4} - \frac{8\pi}{4}$$
$$= \frac{15\pi}{4}$$

That's still too big (the largest $\frac{\pi}{4}$ angle I have memorized is $\frac{7\pi}{4}$), so I have to subtract again:

$$\frac{15\pi}{4} - \frac{8\pi}{4} = \frac{7\pi}{4}$$

Because $\frac{7\pi}{4}$ and $\frac{23\pi}{4}$ are coterminal, $\cos \frac{7\pi}{4} = \cos \frac{23\pi}{4} = \frac{\sqrt{2}}{2}$.

> **YOU'VE GOT PROBLEMS**
>
> Problem 1: Evaluate $\cos \frac{14\pi}{4}$ using a coterminal angle and the unit circle.

You might be interested in how the unit circle originates and how the previous values are derived. Here's a quick explanation based on Figure 4.9. A unit circle is just a circle with radius 1, and we'll center it at the origin. Now, draw a segment from the origin that makes a 30-degree $\left(\frac{\pi}{6} \text{ radian}\right)$ angle with the positive *x*-axis in the first quadrant, and mark the point where the ray intersects the circle. The coordinates of that point are, respectively, the cosine and sine of $\frac{\pi}{6}$. To find the coordinates of the point, find the lengths of the legs of the right triangle.

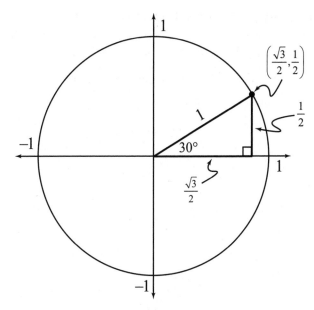

Figure 4.9
The 30-60-90 right triangle here has legs whose lengths are the cosine and sine value
of $30° = \frac{\pi}{6}$ *, the angle at the origin.*

Incredibly Important Identities

An identity is an equation that is always true, regardless of the input. It's easy to tell that, according to this definition, $x + 1 = 7$ is not an identity, because it is only true when $x = 6$. However, consider the equation $2(x - 1) + 3 = 2x + 1$. If you plug in $x = 0$, you get $1 = 1$, which is definitely true, wouldn't you agree? Try plugging in any real number, and you'll get another true statement. Thus, $2(x - 1) + 3 = 2x + 1$ is an identity.

It is worth mentioning that it is a very stupid identity. You are not going to impress anyone by showing that equation off. With only two seconds' worth of work, you can simplify the left side and show that the two sides are equal. Most math identities are much more useful because it's not immediately obvious that they are true. Specifically, trigonometric identities help you rewrite equations, simplify expressions, and justify answers to equations. With this in mind, we'll explore the most common trigonometric identities. They're worth memorizing if you don't know them already.

Pythagorean Identities

The three most important of all the trig identities are the Pythagorean identities. They are named as such because they are created with the Pythagorean theorem. Remember that little nugget from geometry? It said that the sum of the squares of the legs of a right triangle is equal to the square of the hypotenuse: $a^2 + b^2 = c^2$. I have always called these identities the Mama, Papa, and Baby theorems (after the Three Bears story), both for entertainment purposes and because they have no commonly accepted names. If I were in the company of math nerds, however, I wouldn't use the terms. They will sneer and scowl at you.

- Mama theorem: $\cos^2 x + \sin^2 x = 1$

- Papa theorem: $1 + \tan^2 x = \sec^2 x$

- Baby theorem: $1 + \cot^2 x = \csc^2 x$

YOU'VE GOT PROBLEMS

Problem 2: The Mama theorem is an identity, and therefore true for every input. Show that it is true for $x = \frac{2\pi}{3}$.

The Wizard of Oz fans out there may remember that the Scarecrow spouts a formula when the Great and Powerful Oz grants him a brain. He states, "The sum of the square roots of any two sides of an isosceles triangle is equal to the square root of the remaining side." This is a false statement! He was probably supposed to quote the Pythagorean theorem, since it is one of the most recognizable theorems to the general public, but missed it quite badly. Perhaps Oz was not so powerful after all. The Tin Man's string of failed marriages and the Cowardly Lion's lack of success as a motivational speaker offer further evidence to Oz's lackluster gift-giving.

Now, let's see how trigonometric functions and identities can make our lives easier. With a little knowledge of trigonometry and a little bit of elbow grease, even ugly expressions can be made beautiful.

Example 4: Simplify the trigonometric expression $\frac{\cos^2 x}{\sin x} + \sin x$ using a Pythagorean identity.

Solution: One of these terms is a fraction. You know that you must first have common denominators in order to add fractions, so multiply the second term by $\frac{\sin x}{\sin x}$ to get his-and-hers matching denominators of $\sin x$:

$$\frac{\cos^2 x}{\sin x} + \frac{\sin x}{1} \cdot \frac{\sin x}{\sin x} = \frac{\cos^2 x}{\sin x} + \frac{\sin^2 x}{\sin x}$$

Now that the denominators match, you can perform the addition in the numerator while leaving the denominator alone:

$$\frac{\cos^2 x + \sin^2 x}{\sin x}$$

KELLEY'S CAUTIONS

The notation $\cos^2 x$ is shorthand notation for $(\cos x)^2$. It wouldn't make any sense for the letters *cos* to be squared. The shorthand notation is used to avoid having to write those extra parentheses.

That doesn't look any easier! Hold on a second. The numerator looks just like the Mama theorem, and according to the Mama theorem, $\cos^2 x + \sin^2 x = 1$. Therefore, substitute 1 in for the numerator:

$$\frac{1}{\sin x}$$

You could stop there, but you're on a roll! You also know that $\frac{1}{\sin x} = \csc x$ since the cosecant is the reciprocal of the sine. Therefore, the final answer is $\csc x$.

Double-Angle Formulas

These identities allow you to write trigonometric expressions containing double angles (such as $\sin 2x$ and $\cos 2x$) into equivalent single-angle expressions. In other words, these expressions eliminate a 2 coefficient inside a trigonometric expression.

- $\sin 2x = 2\sin x \cos x$ (This is the simplest double-angle formula, and memorizing it is a snap.)

- $\cos 2x = \cos^2 x - \sin^2 x$
  $$= 2\cos^2 x - 1$$
  $$= 1 - 2\sin^2 x$$

The cosine double-angle formulas are a little trickier—there are actually three different things that can be substituted for $\cos 2x$. You should choose which to substitute in based on the rest of the problem. If there seem to be a lot of sines in the equation or expression, use the last of the three, for example.

There isn't a whole lot to understand about double-angle formulas. You should just be ready to recognize them at a moment's notice, as problems very rarely contain the warning label, "Caution: This problem will require you to know basic trig double-angle formulas. Keep away from eyes. May pose a choking hazard to children under 3." Watch how slyly these suckers slip in there.

CRITICAL POINT

There are a lot of trig identities—not just the few in this chapter—but you'll use these far more than all the rest put together.

Example 5: Factor and simplify the expression $\cos^4 \theta - \sin^4 \theta$.

Solution: This expression is the difference of perfect squares, so it can be factored as follows: $(\cos^2 \theta + \sin^2 \theta)(\cos^2 \theta - \sin^2 \theta)$. Notice that the left-hand quantity is equal to 1, according to the Mama theorem, and the right-hand quantity is equal to $\cos 2x$, according to our double-angle formulas. Therefore, we can substitute those values to get $(1)(\cos 2x) = \cos 2x$.

YOU'VE GOT PROBLEMS

Problem 3: Factor and simplify the expression $2\sin x \cos x - 4\sin^3 x \cos x$.

Solving Trigonometric Equations

The last really important trig skill you need to possess is the ability to solve trigonometric equations. A word of warning: some math teachers get very bent out of shape when discussing trig equations. You will have to read the directions to these sorts of problems very carefully to make sure to answer the exact question being asked of you. This includes the interval for the solution.

When intervals are specified as [0,2π), that is shorthand for $0 \leq x < 2\pi$. The two numbers in the notation represent the lower and higher boundaries of the acceptable interval, and the bracket or parenthesis tells you whether that boundary is included in the interval or not. If it's a bracket, that boundary is included, but not so with a parenthesis.

CRITICAL POINT

In interval notation, the expression $x \geq 7$ looks like [7,∞). Because there is no upper bound, you write infinity. If infinity is one of the boundaries, you always use a parenthesis next to it.

Each of my examples will ask for the solution to the trigonometric equation on the interval [0,2π). Therefore, there may be multiple answers. Some instructors will demand that you write the specific correct answer for each equation. In other words, although there may be many angles that solve the problem, they only accept one answer. This answer falls within a specific range, and as long as you learn the appropriate range for each trigonometric function, you'll be okay.

The best approach is to ask if they'll require answers on a certain interval or if they expect only the answer on the appropriate range.

The procedure for solving trigonometric equations is not unlike solving regular equations. However, the final step often requires you to remember the unit circle!

> **KELLEY'S CAUTIONS**
>
> If your instructor demands one answer per equation, eliminate all of your solutions except for the one (and there will only be one) that falls into the appropriate range. That range is $-\frac{\pi}{2} \le \theta \le \frac{\pi}{2}$ for sine, tangent, and cosecant; for cosine, cotangent, and secant, use the interval of $0 \le \theta \le \pi$.

Example 6: Solve the equation $\cos 2x - \cos x = 0$ on the interval $[0, 2\pi)$.

Solution: First of all, you want to eliminate the double-angle formula so that all of the terms are single angles. Because you are replacing $\cos 2x$, there are three options, but I will choose the $2\cos^2 x - 1$ option since the problem also contains another cosine term:

$$\left(2\cos^2 x - 1\right) - \cos x = 0$$
$$2\cos^2 x - \cos x - 1 = 0$$

Now you can factor this equation. (If you're having trouble, think of the equation as $2w^2 - w - 1$ and factor that, substituting in $w = \cos x$ when you're finished.)

$$(2\cos x + 1)(\cos x - 1) = 0$$

Like any other quadratic equation solved using the factoring method, set each factor equal to 0 and solve:

$$2\cos x + 1 = 0 \qquad \text{and} \qquad \cos x - 1 = 0$$
$$\cos x = \tfrac{1}{2} \qquad\qquad\qquad \cos x = 1$$

To finish, ask yourself, "When is the cosine equal to $-\frac{1}{2}$ and when is it 1?" The question asks for all answers on $[0, 2\pi)$, so give all the correct answers on the unit circle:

$$x = \tfrac{2\pi}{3}, \tfrac{4\pi}{3}, \text{ or } 0$$

All three answers should be given. If your instructor requires only answers on the appropriate ranges, your solutions would be $x = \frac{2\pi}{3}$ and $x = 0$. You'd throw out $x = \frac{4\pi}{3}$ because it does not fall in the correct cosine range of $0 \le \theta \le \pi$. In this case, it's okay to have a total of two answers, since each of the individual, smaller equations has one answer.

> **YOU'VE GOT PROBLEMS**
>
> Problem 4: Solve the equation $\sin 2x + 2\sin x = 0$ and provide all solutions on the interval $[0,2\pi)$.

The Least You Need to Know

- The six basic trigonometric functions are sine, cosine, tangent, cotangent, secant, and cosecant.
- Sine and cosine's values are used to evaluate the other four trig functions.
- There are some angles on the interval $[0,2\pi)$ whose cosine and sine values you should have memorized.
- Trigonometric identities help you simplify trig expressions and solve trig equations.

Laying the Foundation for Calculus

I have some good news and some bad news. The good news is that we'll be dealing with relatively easy functions for the remainder of calculus. The bad news is that we need to mathematically define exactly what we mean by "easy." When math people sit down to define things, you know that theorems are going to start flying around, and calculus is no different.

When we say "easy" functions, we really mean continuous ones. In order to be continuous, the function can't contain holes and isn't allowed to have any breaks in it. That sounds nice, but math people like their definitions more specific (read: complicated) than that. In order to define "continuous," we'll first need to design something called a "limit." During this part of the book, you'll learn what a limit is, how to evaluate limits for functions, and how to apply limits to design a definition for continuity.

CHAPTER 5

Take It to the Limit

When most people look back on calculus after completing it, they wonder why they had to learn limits at all. For some, it's like getting all of their teeth pulled just for the fun of it. After a brief limit discussion at the start of the course, there are very few times that limits return, and when they do, it is only for a brief cameo role in the topic at hand. However, limits are extremely important in the development of calculus and in all of the major calculus techniques, including differentiation, integration, and infinite series.

As I discussed in Chapter 1, limits were the key ingredient in the discovery of calculus. They allow you to do things that ordinary math gets cranky about. In practice, limits are many students' first encounter with a slightly philosophical math topic, answering questions like, "Even though this function is undefined at this *x*-value, what height did it *intend* to reach?" This chapter will give you a great intuitive feel of what a limit is and what it means for a function to have a limit; the next chapter will help you evaluate limits.

One final note: the official limit definition is called the delta-epsilon definition of limits. It is very complex, and is based on high-level mathematics. A discussion of this rigorous mathematical concept is not beneficial, so it is omitted here. In essence, it is possible to be a great driver without having to understand every principle of the combustion engine.

In This Chapter

- Understanding what a limit is
- Why limits are needed
- Approximating limits
- One-sided and general limits

What Is a Limit?

When I first took calculus in high school, I was hip-deep in evaluating limits via tons of different techniques before I realized that I had no idea what I was doing, or why. I am one of those people who needs some sort of universal understanding in a math class, some sort of framework to visualize why I am undertaking the process at hand. Unfortunately, calculus teachers are notorious for explaining *how* to complete a problem (outlining the steps and rules) but not explaining *what the problem means.* So for your benefit and mine, we'll discuss what a limit actually is before we get too nutty with the math part of things.

Let's start with a simple function: $f(x) = 2x + 5$. You know that this is a line with slope 2 and y-intercept 5. If you plug $x = 3$ into the function, the output will be $f(3) = 2(3) + 5 = 11$. Very simple, everyone understands, everyone's happy. What else does this mean, however? It means that the point (3,11) belongs to the relation and function I call f. Furthermore, it means that the point (3,11) falls on the graph of $f(x)$, as evidenced in Figure 5.1.

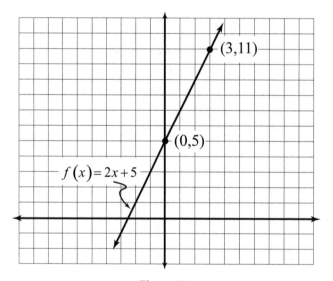

Figure 5.1
The point (3,11) falls on the graph of f(x).

All of this seems pretty obvious, but let's change the way we talk just a little to prepare for limits. Notice that as you get closer and closer to $x = 3$, the height of the graph gets closer and closer to $y = 11$. In fact, if you plug $x = 2.9$ into $f(x)$, you get $f(2.9) = 2(2.9) + 5 = 10.8$. If you plug in $x = 2.95$, the output is 10.9. Inputs close to 3 give outputs close to 11, and the closer the input is to 3, the closer the output is to 11.

Even if you didn't know that $f(3) = 11$ (say for some reason you were forbidden by your evil stepmother, as was Cinderella), you could still figure out what it would *probably* be by plugging in an insanely close number like 2.99999. I'll save you the grunt work and tell you that $f(2.99999) =$ 10.99998. It's pretty obvious that f is headed straight for the point (3,11), and that's what is meant by a limit.

A *limit* is the intended height of a function at a given value of x, whether or not the function actually reaches that height at the given x. In the case of f, you know that f does reach the value of 11 when $x = 3$, but that doesn't have to be the case for a limit to exist. Remember that a limit is the height a function *intends* to reach.

 **DEFINITION**

A **limit** is the height a function *intends* to reach at a given x value, whether or not it actually reaches it.

Can Something Be Nothing?

You may ask, "How am I supposed to know what a function *intends* to do? I don't even know what *I* intend to do." Luckily, functions are a little more predictable than people, but more on that later. For now, let's look at a slightly harder problem involving limits. But before we do, let's discuss how a limit is written in calculus.

In our previous example, we determined that the limit, as x approaches 3, of $f(x)$ equals 11, because the function approached a height of 11 as we plugged in x values closer and closer to 3. As it seems with everything else, calculus has a shorthand notation for this:

$$\lim_{x \to 3} f(x) = 11$$

This is read, "The limit, as x approaches 3, of $f(x)$ equals 11." The tiny 3 is the number you're approaching, $f(x)$ is the function in question, and 11 is the intended height of f at 3. Now, let's look at a slightly more involved example.

Figure 5.2 is the graph of $g(x) = \frac{x^2 - x - 6}{x + 2}$. Clearly, the domain of g cannot contain $x = -2$, because that causes 0 in the denominator, and that is just plain yucky.

Notice that the graph of g has a hole at the evil value of $x = -2$, but that won't stop us. We're going to evaluate the limit there. Remember, the function doesn't actually have to exist at a certain point for a limit to exist—the function only has to have a clear height it intends to reach. Clearly, the function has an intended height it wishes to reach when $x = -2$ in the graph—there's a gaping hole at that exact spot, in fact.

CRITICAL POINT

If you substitute $x = -2$ into $g(x) = \frac{x^2-x-6}{x+2}$, you get $\frac{0}{0}$, which is said to be in "indeterminate form." Typically, a result of $\frac{0}{0}$ means that a hole appears in the graph at that value of x, which is the case with $g(x)$.

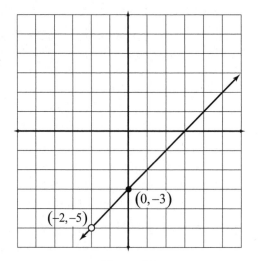

Figure 5.2

The graph of $g(x) = \frac{x^2-x-6}{x+2}$.

CRITICAL POINT

We will evaluate limits like $\lim\limits_{x \to 3} f(x)$ and $\lim\limits_{x \to -2} g(x)$ in the next chapter without having to resort to the "plug in an insanely close number" technique. In this chapter, focus with me on the idea of a limit, and we'll get to the computational part soon enough.

How can you evaluate $\lim\limits_{x \to -2} g(x)$? Just as we did in the previous example, you'll plug in a number insanely close to $x = -2$, in this case, $x = -1.99999$. Again, I'll do the grunt work for you (you can thank me later): $g(-1.99999) = -4.99999$. Even a knucklehead like me can see that this function intends to go to a height of -5 on the function g when $x = -2$.

Therefore, $\lim\limits_{x \to -2} g(x) = -5$, even though the point $(-2,-5)$ does not appear on the graph of $g(x)$. This is one example of a limit existing because a function intends to go to a height despite not actually reaching that height.

Example 1: Graph $f(x) = \frac{2x^2-9x+4}{x-4}$ and simplify the function to evaluate $\lim\limits_{x \to 4} f(x)$.

Solution: You might be thinking, "Graph something that complicated? How am I supposed to do that?" While you could spend an hour plotting points on the graph by substituting x-values into the function, there's no need. Like the function $g(x)$ graphed in Figure 5.2, $f(x)$ will have a much simpler graph than you may initially think.

Begin by factoring the numerator of $f(x)$: $f\left(x\right)=\frac{(x-4)(2x-1)}{x-4}$

Notice that the common factor $(x - 4)$ appears in the numerator and denominator of the fraction. You can simplify the fraction by eliminating the common factor, basically crossing out $(x - 4)$ like so:

$$f\left(x\right)=\frac{\cancel{(x-4)}(2x-1)}{\cancel{x-4}}$$
$$f\left(x\right)=2x-1$$

This means the values of $f\left(x\right)=\frac{2x^2-9x+4}{x-4}$ are exactly the same as the values of $f(x) = 2x - 1$, with one gigantic exception. The original version of the function is a fraction, and if you substitute $x = 4$ into it, you get 0 in the denominator, which is not allowed.

What does all that mean? The graphs of $f\left(x\right)=\frac{2x^2-9x+4}{x-4}$ and $f(x) = 2x - 1$ look exactly the same except when $x = 4$. At that point on the line, you should place a hole in the graph, as illustrated in Figure 5.3.

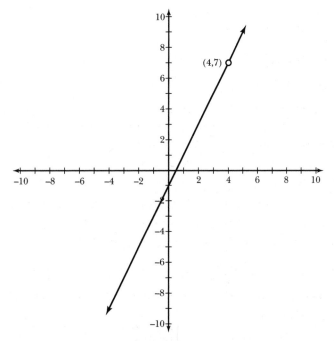

Figure 5.3
The graph of $f\left(x\right)=\frac{2x^2-9x+4}{x-4}$, which is not defined when x = 4. *Note that it matches the graph of* y = 2x – 1, *except at* x = 4.

Now to calculate $\lim\limits_{x \to 4} f(x)$. Even though $f(x)$ is not defined when $x = 4$, $y = 2x - 1$ is, and we know that the graphs exactly match each other everywhere else. Therefore, $f(x) = \frac{2x^2 - 9x + 4}{x - 4}$ *intends* to reach the height $f(x) = 2x - 1$ reaches when $x = 4$. To calculate that height, substitute $x = 4$ into the simplified version of $f(x)$.

$$f(4) = 2(4) - 1$$
$$= 8 - 1$$
$$= 7$$

You conclude that $\lim\limits_{x \to 4} f(x) = 7$, which you can verify visually using the graph in Figure 5.3.

> **YOU'VE GOT PROBLEMS**
>
> Problem 1: Graph $j(x) = \frac{x^2 + 2x - 15}{x - 3}$ and evaluate $\lim\limits_{x \to 3} j(x)$.

One-Sided Limits

Occasionally, a function will intend to reach two different heights at a given x, one height as you come from the left side and one height as you come from the right side. We can still describe these one-sided intended heights, using *left-hand* and *right-hand limits*. To better understand this bizarre function behavior, look at the graph of $h(x)$ in Figure 5.4.

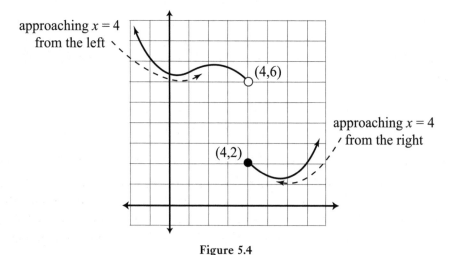

Figure 5.4

The graph of h(x) *consists of two pieces; a graph like this is usually the result of a piecewise-defined function.*

DEFINITION

> A **left-hand limit** is the height a function intends to reach as you approach the given x value *from* the left; the **right-hand limit** is the intended height as you approach *from* the right.

This graph does something very wacky at $x = 4$: it breaks. Trace your finger along the graph as it approaches $x = 4$ from the left. What height is your finger approaching as you get close to (but don't necessarily reach) $x = 4$? You are approaching a height of 6. This is called the left-hand limit and is written like this:

$$\lim_{x \to 4^-} h(x) = 6$$

CRITICAL POINT

> To keep from confusing right- and left-hand limits, remember the key word: *from*. A left-hand limit is the height toward which you're heading as you approach the given x-value from the left, not as you go *toward* the left on the graph.

The little negative sign in the exponent indicates that you should only be interested in the height the graph approaches as you travel along the graph from the left-hand side. If you trace your finger along the other portion of the graph, this time toward $x = 4$ from the right, you'll notice that you approach a height of 2 when you get close to $x = 4$. This is, as you may have guessed, the right-hand limit for $x = 4$, and it is written as follows:

$$\lim_{x \to 4^+} h(x) = 2$$

Example 2: Graph the function $k(x)$, defined here. Then, evaluate (a) $\lim_{x \to -1^-} k(x)$ and (b) $\lim_{x \to -1^+} k(x)$.

$$k(x) = \begin{cases} -x - 3, & -1 > x \geq 1 \\ 2x + 2, & -1 \leq x < 1 \end{cases}$$

Solution: If you need to review piecewise-defined functions, check out Example 2 in Chapter 3.

The function $k(x)$ is defined by two linear equations. Its values come from $y = -x - 3$ whenever $-1 > x \geq 1$. It may help you to split that compound inequality into two simple inequalities: $-1 > x$ and $x \geq 1$. (It also may help to rewrite $-1 > x$ with x on the left side: $x < -1$.) Basically, values of $k(x)$ are generated by the expression $-x - 3$ for x-values less than -1 or greater than or equal to 1.

Similarly, values of $k(x)$ are generated by the expression $2x + 2$ for x-values greater than or equal to -1 and less than 1. (Break the compound inequality into two simple inequalities again if that helps you visualize the interval.) The graph of $k(x)$ appears in Figure 5.5.

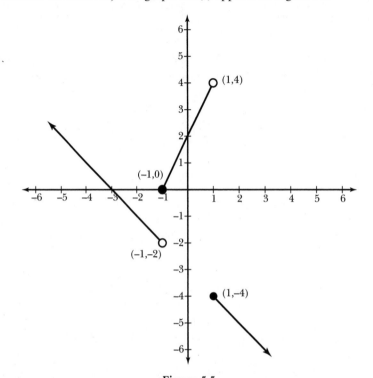

Figure 5.5
The graph of k(x) *and four points of interest, the endpoints of the linear segments that form* k(x)*.*

Although your graph may be slightly inaccurate if you mix up the open and solid dots on your graph, it won't affect the limit values at $x = -1$ and $x = 1$. Speaking of the limit values, it's time to get calculating.

To evaluate $\lim\limits_{x \to -1^-} k(x)$, you need to identify the height that $k(x)$ *intends* to reach as you approach $x = -1$ from the left. Trace your finger from left to right along the graph, beginning at its left edge and approaching $x = -1$. Stop at $x = -1$, right before the graph jumps from point $(-1,-2)$ to $(-1,0)$. As you approach $x = -1$ from the left, the function intends to reach point $(-1,-2)$, with an intended height of -2. Therefore, $\lim\limits_{x \to -1^-} k(x) = -2$.

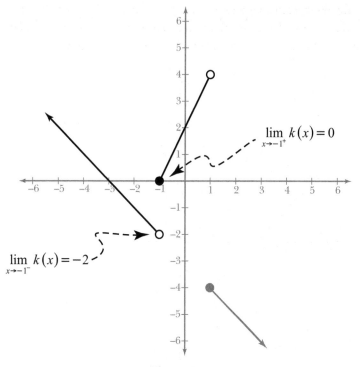

Figure 5.6
Calculating the left- and right-hand limits of k(x) *as* x *approaches −1.*

To calculate $\lim\limits_{x \to -1^+} k(x)$, approach that same break in the graph at $x = -1$, but this time approach from the right side, tracing your finger along $y = 2x + 2$ as it declines steeply. (Remember, you're approaching $x = -1$ from right to left.) Again, stop before the jump in the graph at $x = -1$, this time at the point $(-1,0)$. This line has an intended function height of 0, so $\lim\limits_{x \to -1^+} k(x) = 0$.

YOU'VE GOT PROBLEMS

Problem 2: Using the piecewise-defined function $k(x)$ defined in Example 2, calculate $\lim\limits_{x \to 1^-} k(x)$ and $\lim\limits_{x \to 1^+} k(x)$.

Until now, we have only spoken of a general limit (in other words, a limit that doesn't involve a direction, such as from the right or left). Most of the time in calculus, you will worry about general limits, but in order for general limits to exist, right- and left-hand limits must also be present; this we learn in the next section, which will tie together everything we've discussed so far about limits. Can you feel the electricity in the air?

When Does a Limit Exist?

If you don't understand anything else in this chapter, make sure to understand this section. It contains the two essential characteristics of limits: when they exist and when they don't. If you've understood everything so far, you're on the verge of understanding your first major calculus topic. I'm so proud of you—I remember when you were only *this* tall.

Here's the key to limits: in order for a limit to exist on a function f at some x-value (we'll give it a generic name like $x = c$), three things must happen:

1. The left-hand limit must exist at $x = c$.

2. The right-hand limit must exist at $x = c$.

3. The left- and right-hand limits at c must be equal.

In calculus books, this is usually written like this: if $\lim\limits_{x \to c^-} f(x) = \lim\limits_{x \to c^+} f(x)$, then $\lim\limits_{x \to c} f(x)$ exists and is equal to the one-sided limits.

The diagram in Figure 5.7 will help illustrate the point.

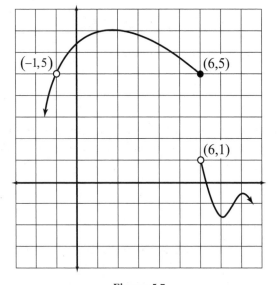

Figure 5.7
Yet another hideous graph called f(x). *Can you spot where the limit doesn't exist?*

There are two interesting x-values on this graph: $x = -1$ and $x = 6$. At one of those values, a general limit exists, and at the other, no general limit exists. Can you figure out which is which using the guidelines?

You're reading ahead, aren't you? Well, stop it. Don't read any more until you've actually tried to answer the question I've asked you. Do it. I'm watching!

The answer: $\lim_{x \to -1} f(x)$ exists and $\lim_{x \to 6} f(x)$ does not. Remember, in order for a limit to exist, the left- and right-hand limits must exist at that point and be equal. As you approach $x = -1$ from the left and right sides, each time you are heading toward a height of 5, so the two one-sided limits exist and are equal, and we can conclude that $\lim_{x \to -1} f(x) = 5$ (i.e., the general limit as x approaches -1 on $f(x)$ is equal to 5).

However, this is not the case when we approach $x = 6$ from the right and left. In fact, $\lim_{x \to 6^-} f(x) = 5$, whereas $\lim_{x \to 6^+} f(x) = 1$. Because those one-sided limits are unequal, we say that no general limit exists at $x = 6$, and that $\lim_{x \to 6} f(x)$ does not exist.

Visually, a limit exists if the graph does not break at that point. For the graph $f(x)$ in question, a break occurs at $x = 6$ but not $x = -1$, which means a limit doesn't exist at the break but can exist at the hole in the graph. Remember that a limit can exist even if the function doesn't exist there—as long as the function intends to reach the same height from the left and the right, the limit exists.

When Does a Limit Not Exist?

You already know of one instance in which limits don't exist, but two other circumstances can ruin a limit as well.

- A general limit does not exist if the left- and right-hand limits aren't equal.

In other words, if there is a break in the graph of a function, and the two pieces of the function don't meet at an intended height, then no general limit exists there. In Figure 5.8, $\lim_{x \to c} g(x)$ does not exist because the left- and right-hand limits are unequal.

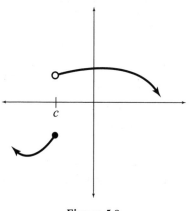

Figure 5.8
The graph of g(x).

- A general limit does not exist if a function increases or decreases infinitely at a given x-value (i.e., the function increases or decreases without bound).

In order for a general limit to exist, the function must approach some fixed numerical height. If a function increases or decreases infinitely, then no limit exists. In Figure 5.9, $\lim_{x \to c} h(x)$ does not exist because $h(x)$ has a vertical asymptote at $x = c$, causing the function to increase without bound there. A limit must be a finite number in order to truly exist.

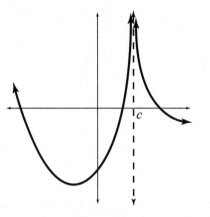

Figure 5.9
The graph of h(x)*.*

- A general limit does not exist if a function oscillates infinitely, never approaching a single height.

This is rare, but sometimes a function will continually wiggle back and forth, never reaching a single numeric value. If this is the case, then no general limit exists. Because this is so rare, most calculus books give the same example when discussing this eventuality, and I will be no different. (Math peer pressure is harsh, let me tell you.) No general limit exists at $x = 0$ in Figure 5.10 because the function never settles on any one value the closer you get to $x = 0$.

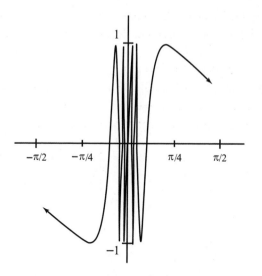

Figure 5.10

The graph of $y = \sin \frac{1}{x}$; $\lim\limits_{x \to 0} \sin \frac{1}{x}$ does not exist.

Example 3: A function $f(x)$ is defined by the graph in Figure 5.11. Based on the graph and your amazing knowledge of limits, evaluate the limits that follow. If no limit exists, explain why.

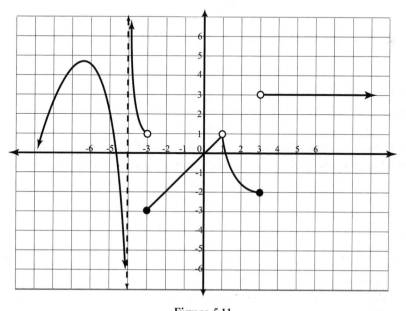

Figure 5.11

The hypnotic graph of $f(x)$. Mortal men may turn to stone upon encountering its terrible visage.

(a) $\lim\limits_{x \to -4^+} f(x)$

Solution: As you approach $x = -4$ from the right, the function increases without bound. You have two ways to write your answer; either say that the limit does not exist because the function increases infinitely, or write $\lim\limits_{x \to -4^+} f(x) = \infty$.

(b) $\lim\limits_{x \to 4} f(x)$

Solution: As you approach $x = 4$ from the right and left, the function approaches a height of 3. Therefore, the general limit exists and is 3.

(c) $\lim\limits_{x \to -3} f(x)$

KELLEY'S CAUTIONS

If a graph has no general limit at one of its x-values, that does not affect any of the other x-values. For example, in Figure 5.10, a general limit exists at every x-value except $x = 0$.

Solution: No general limit exists here because the left-hand limit (1) does not equal the right-hand limit (−3).

CRITICAL POINT

Giving a limit answer of ∞ or $-\infty$ is equivalent to saying that the limit does not exist. However, by answering with ∞, you are also explaining why the limit doesn't exist and specifically detailing whether the function increased or decreased infinitely there.

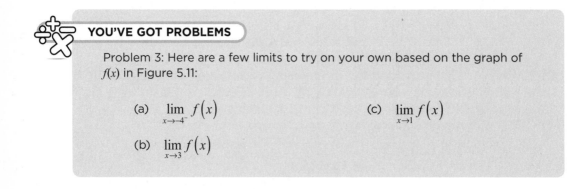

YOU'VE GOT PROBLEMS

Problem 3: Here are a few limits to try on your own based on the graph of $f(x)$ in Figure 5.11:

(a) $\lim\limits_{x \to -4^-} f(x)$ (c) $\lim\limits_{x \to 1} f(x)$

(b) $\lim\limits_{x \to 3} f(x)$

The Least You Need to Know

- The limit of a function at a given x-value is the height the function intends to reach there.

- A function can have a limit at an x-value even if the function has a hole there.

- A function cannot have a limit where its graph breaks.

- If a function's left- and right-hand limits exist and are equal for a certain $x = c$, then a general limit exists at c.

- A limit does not exist in the cases of infinite function growth or oscillation.

CHAPTER 6

Evaluating Limits Numerically

Now you know what a limit is, when a limit exists, and when it doesn't. However, the question of how to actually evaluate limits remains. In Chapter 5 we approximated limits by plugging in x values insanely close to the number we were approaching, but that got tedious quickly. As soon as you have to raise numbers like 2.999999 to various exponents, it becomes clear that you either need a better way or a giant bottle of aspirin.

Good news: there are lots of better ways, and this chapter will lead you through all of the major processes to evaluate limits and the important limit theorems you should memorize. For those of you who were uncomfortable with math turning a little conceptual and philosophical there for a little bit, don't worry—we're back to comfortable, familiar, soft, fuzzy, and predictable math techniques and formulas.

All of that theory you learned in the last chapter will resurface to some degree in Chapter 7, when we discuss continuity of functions, so keep it fresh in your mind. A lot of our discussion about limits will get hazy quickly when you move on to derivatives and integrals as the book progresses. Make sure you come back and review these early topics often throughout your calculus course to keep them fresh in your brain.

In This Chapter

* Three easy methods for finding limits
* Limits and asymptotes
* Finding limits at infinity
* Trig and exponential limit theorems

The Major Methods

The vast majority of limits can be evaluated by using one of three techniques: substitution, factoring, and the conjugate method. Usually, only one of these techniques will work on a given limit problem, so you should try one method at a time until you find one that works. Because I am efficient (understand, by that I mean extremely lazy) I always try the easiest method first, and only move on to more complicated methods if I absolutely have to. As such, I'll present the methods from easiest to hardest.

Substitution Method

Prepare yourself—you're going to weep with uncontrollable joy when I tell you this. Many limits can be evaluated simply by plugging the x value you're approaching into the function. The fancy term for this is the *substitution method* (or the *direct substitution method*).

Example 1: Evaluate $\lim_{x \to 4} \left(x^2 - x + 2 \right)$.

Solution: In order to evaluate the limit, simply plug the number you're approaching (4) in for the variable:

$$4^2 - 4 + 2 = 16 - 2 = 14$$

According to the substitution method, $\lim_{x \to 4} \left(x^2 - x + 2 \right) = 14$. That was too easy! Just to make sure it actually worked, let's check the answer by looking at the graph of $y = x^2 - x + 2$ in Figure 6.1.

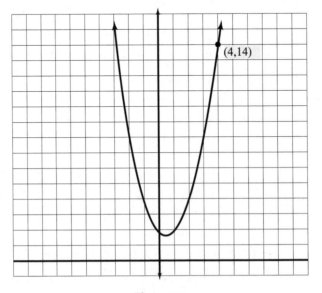

Figure 6.1

Use the graph of y + x² − x + 2 *to visually verify the limit at* x = 4.

As we approach $x = 4$ from either the left or the right, the function clearly heads toward a height of 14, which, as we know, guarantees that the general limit exists and is 14. It worked! Huzzah!

> ✏️ **CRITICAL POINT**
>
> When I say that the methods of evaluating limits are listed from easiest to hardest, I should qualify it by saying that *hard* is not a good word choice; none of these methods is hard. The number of steps increases slightly from one method to the next, but these methods are easy.

If every limit problem in the world could be solved using substitution, there would probably be no need for mathematically induced antidepressants. However (and there's always a however, isn't there?), sometimes substitution cannot be used. In such cases, you should resort to the next method of evaluating limits: factoring.

> ✖️➕ **YOU'VE GOT PROBLEMS**
>
> Problem 1: Evaluate the following limits using substitution:
>
> (a) $\lim\limits_{x \to \pi} \frac{\cos x}{x}$
>
> (b) $\lim\limits_{x \to -2} \frac{x^2+1}{x^2-1}$

Factoring Method

Consider the function $f(x) = \frac{x^2-9}{x+3}$. How would you find the limit of $f(x)$ as x approaches -3? Well, if you try to use substitution to find the limit, bad things happen:

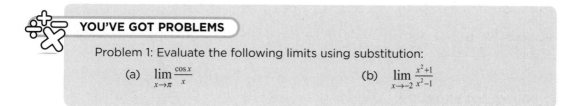

$$\lim_{x \to -3} \frac{x^2-9}{x+3} = \frac{(-3)^2-9}{-3+3}$$
$$= \frac{9-9}{0}$$
$$= \frac{0}{0}$$

What kind of an answer is $\frac{0}{0}$? A gross one, that's for sure. Remember that we can't have 0 in the denominator of a fraction; that's not allowed. If you recall, in the last chapter we called this an "indeterminate form," which means the answer can be anything. Clearly, then, the limit is not $\frac{0}{0}$, but that answer does tell us two things:

1. You must use a different method to find the limit, because …

2. … the function likely has a hole at the x value you substituted into the function, and you need to determine the height of that hole.

The best alternative to substitution is the factoring method, which works just beautifully in this case. In the next example, we'll find this troubling limit.

Example 2: Evaluate $\lim\limits_{x \to -3} \frac{x^2-9}{x+3}$ using the factoring method.

Solution: To begin the factoring method, factor! It makes sense, because the numerator is the difference of perfect squares and factors very happily:

$$\lim_{x \to -3} \frac{(x+3)(x-3)}{x+3}$$

Now both the top and bottom of the fraction contain $(x + 3)$, so you can cancel those terms out to get the much simpler limit expression of:

$$\lim_{x \to -3} (x-3)$$

Now you can use the substitution method to finish:

$$-3 - 3 = -6$$

So $\lim\limits_{x \to -3} \frac{x^2-9}{x+3} = -6$.

YOU'VE GOT PROBLEMS

Problem 2: Evaluate these limits using the factoring method:

(a) $\lim\limits_{x \to 5} \frac{2x^2-7x-15}{x-5}$

(b) $\lim\limits_{x \to 1} \frac{x^3-1}{x-1}$

Conjugate Method

If substitution and factoring don't work, you have one last bastion of hope when evaluating limits, but this final method is very limited in its scope and power. In fact, it is most useful for limits that contain radicals, as its power comes from the use of the conjugate. The *conjugate* of a binomial expression (i.e., an expression with two terms) is the same expression with the opposite middle sign. For example, the conjugate of $\sqrt{x}-5$ is $\sqrt{x}+5$.

DEFINITION

For our purposes, the **conjugate** of a binomial expression simply changes the sign between the two terms to its opposite. For example, $3+\sqrt{x}$ and $3-\sqrt{x}$ are conjugates.

The true power of conjugate pairs is displayed when you multiply them together. The product of two conjugates containing radicals will, itself, contain no radical expressions! In other words, multiplying by a conjugate can eliminate square roots:

$$\left(\sqrt{x}-5\right)\left(\sqrt{x}+5\right)=\sqrt{x^2}+5\sqrt{x}-5\sqrt{x}-25$$
$$=\sqrt{x^2}-25$$
$$=x-25$$

You should use the conjugate method whenever you have a limit problem containing radicals for which substitution does not work—always try substitution first. However, if substitution results in an illegal value $\left(\text{like } \frac{0}{0}\right)$, you'll know to employ the conjugate method, which we'll use to solve the next example.

Example 3: Evaluate $\lim\limits_{x\to5}\frac{\sqrt{x+11}-4}{x-5}$.

Solution: If you try the substitution method, you get $\frac{0}{0}$, indicating that you'll need another method to find the limit because the function probably has a hole at $x = 5$. The function itself contains a radical and a number being subtracted from it—the fingerprint of a problem needing the conjugate method. To start, multiply both the numerator and denominator by the conjugate of the radical expression $\left(\sqrt{x+11}+4\right)$:

$$\lim\limits_{x\to5}\frac{\sqrt{x+11}-4}{x-5}\cdot\frac{\sqrt{x+11}+4}{\sqrt{x+11}+4}$$

Multiply the numerators and denominators as you would any pair of binomials—i.e., $(a + b)(c + d) = ac + ad + bc + bd$—and all of the radical expressions will disappear from the numerator. *Do not actually multiply the nonconjugate pair together.* You'll see why in a second:

$$\lim\limits_{x\to5}\frac{\left(\sqrt{x+11}\right)^2+4\sqrt{x+11}-4\sqrt{x+11}-4^2}{(x-5)\left(\sqrt{x+11}+4\right)}$$
$$=\lim\limits_{x\to5}\frac{x+11-16}{(x-5)\left(\sqrt{x+11}+4\right)}$$
$$=\lim\limits_{x\to5}\frac{x-5}{(x-5)\left(\sqrt{x+11}+4\right)}$$

Here's the neat trick: the numerator and denominator now contain the same term $(x - 5)$ so you can cancel that term and then finish the problem with the substitution method:

$$=\frac{1}{\sqrt{x+11}+4}$$
$$=\frac{1}{\sqrt{5+11}+4}$$
$$=\frac{1}{\sqrt{16}+4}$$
$$=\frac{1}{4+4}$$
$$=\frac{1}{8}$$

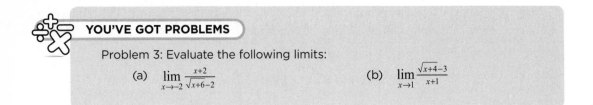

What If Nothing Works?

If none of the three techniques we have discussed works on the problem at hand, you're not out of hope. Don't forget we have an alternative (albeit tedious, mechanical, and unexciting—like most television sitcoms) method of finding limits. If all else fails, substitute a number insanely close to the number for which you are evaluating, as we did in Chapter 5.

Let me also play the part of the soothsayer for a moment. For maximum effect, read the next sentences in a creepy fortune-teller voice. "I see something in your future, yes, off in the distance. A promised method, a shortcut, a new way to evaluate limits that makes hard things easy. I'm getting a French name … L'Hôpital's Rule … and an unlucky number … 13. Chapter 13. Look for it in Chapter 13."

Limits and Infinity

There is a deep relationship between limits and infinity. At first they thought they were "just friends," and then one would occasionally catch the other in a sidelong glance with eyes that spoke volumes. Without going into the long history, now they're inseparable, and without their storybook relationship, there'd be no vertical or horizontal asymptotes.

Vertical Asymptotes

You already know that a limit does not exist if a function increases or decreases infinitely, such as at a vertical asymptote. You may be wondering if it's possible to tell if a function is doing just that without having to draw the graph, and the answer is yes. Just as a substitution result of $\frac{0}{0}$ typically means a hole exists on the graph, a result of $\frac{5}{0}$ indicates a vertical asymptote. To be more specific, you don't have to get 5 in the numerator—any nonzero number divided by 0 indicates that the function is increasing or decreasing without bound, meaning no limit exists.

Example 4: At what value(s) of x does no limit exist for $f(x) = \dfrac{x^2+7x+10}{x^2-25}$?

Solution: Begin by factoring the expression, because knowing what x values cause a 0 in the denominator is key:

$$f(x) = \frac{(x+5)(x+2)}{(x+5)(x-5)}$$

At $x = -5$, the function should have a hole, as substituting in that value results in $\frac{0}{0}$. You can use the factoring method to actually find that limit:

$$\lim_{x \to -5} f(x) = \frac{x+2}{x-5}$$

$$= \frac{-5+2}{-5-5}$$

$$= \frac{-3}{-10}$$

$$= \frac{3}{10}$$

However, you're supposed to determine where the limit *doesn't* exist, so let's look at the other distressing x-value: $x = 5$. If you substitute that into $f(x)$, you get $\frac{70}{0}$. This result, any number (other than 0) divided by 0, indicates the presence of a vertical asymptote at $x = 5$, so $\lim_{x \to 5} f(x)$ does not exist because f will either increase or decrease infinitely there.

If substitution results in $\frac{0}{0}$, that does not *guarantee* that a hole exists in the function. You can only be sure there's a hole there if a limit exists, as was the case with $x = -5$ in this example.

Example 5: Calculate all values of c for which $\lim_{x \to c} \frac{2x^2-6x}{(rx+s)(vx+w)} = \infty$ or $-\infty$.

Solution: This problem looks like a spilled bowl of alphabet soup, doesn't it? Don't let all of the unknowns in there discourage you. Although $r, s, v,$ and w don't look like numbers, that's all they are. You don't know their values, but don't let that bother you. Think of them as little boxes, hidden inside each of which is a plain old harmless number.

Because the limit of the function is either ∞ or $-\infty$, technically that limit doesn't exist because the rational function either increases or decreases without bound near that value of c. This happens at a vertical asymptote.

In other words, this question is actually wondering, "At what x-values does this function have a vertical asymptote?" Begin by setting the denominator equal to zero and solving. Remember, the x-values you're looking for will cause the denominator to equal 0 but *not* the numerator.

$$(rx+s)(vx+w) = 0$$

$$rx+s = 0 \qquad\qquad vx+w = 0$$

$$rx = -s \quad \text{or} \qquad vx = -w$$

$$x = -\frac{s}{r} \qquad\qquad x = -\frac{w}{v}$$

Now you know the denominator equals 0 whenever $c = -\frac{s}{r}$ or $-\frac{w}{v}$. Why use a c there instead of an x? The question wants specific x-values that cause a vertical asymptote, and it names those values c, so you should do the same. One thing left to consider: these strange c-values won't actually represent vertical asymptotes if their values also cause the numerator to equal 0.

Your next step is to find out when the numerator equals 0 to specifically avoid those values. You'll need to set the numerator equal to 0 and solve using the factoring method.

$$2x^2 - 6x = 0$$
$$2x(x-3) = 0$$

$$2x = 0 \qquad x - 3 = 0$$
$$\text{or}$$
$$x = 0 \qquad x = 3$$

Believe it or not, you're finished! The answer isn't all that satisfying, but here goes nothing. The function will increase or decrease without bound for c-values $-\frac{s}{r}$ and $-\frac{w}{v}$, as long as those values are neither equal to 0 nor equal to 3.

YOU'VE GOT PROBLEMS

Problem 4: Determine the x-values at which $g(x) = \frac{2x^3 - 3x^2 + x}{2x^3 + 5x^2 - 3x}$ is undefined. If possible, evaluate the limits as x approaches each of those values.

Horizontal Asymptotes

Vertical asymptotes are caused by a function's values increasing or decreasing infinitely as that function gets closer and closer to a fixed x-value; so, if a function has a vertical asymptote at $x = c$, we can write $\lim\limits_{x \to c} f(x) = \infty$ or $-\infty$. *Horizontal* asymptotes have a lot of the same components, but everything is reversed. One minor exception: a function can cross a horizontal asymptote (but not a vertical asymptote), because horizontal asymptotes only deal with the "end behavior" of the function, the intended heights of the function at the far right and left ends of the graph.

A horizontal asymptote is the height that a function tries to, but cannot, reach as the function's x-values get infinitely positive or negative. In Figure 6.2, $f(x)$ approaches a height of 5 as x gets infinitely positive and a height of -1 as $f(x)$ becomes infinitely negative.

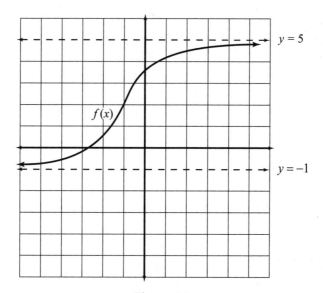

Figure 6.2
The graph of f(x) *has different horizontal asymptotes as* x *gets infinitively positive and negative.*
A rational function won't look like this—it will have (at most) one horizontal asymptote.

This is written as: $\lim_{x\to\infty} f(x) = 5$ and $\lim_{x\to-\infty} f(x) = -1$.

These are called limits *at* infinity, since you are not approaching a fixed number, as you do with typical limits. However, the limit still exists because the function clearly intends to reach the limiting height indicated by the horizontal asymptote, although it never actually reaches it.

Evaluating limits at infinity is a bit different from evaluating standard limits; substitution, factoring, and the conjugate methods won't work, so you need an alternative method. Although L'Hôpital's Rule works quite nicely, you won't learn that until Chapter 13. In the meantime, you can evaluate these limits simply by comparing the highest exponents in their numerators and denominators.

Let's say we calculate $\lim_{x\to\infty} r(x)$, where $r(x)$ is defined as a fraction whose numerator, $n(x)$, and denominator, $d(x)$, are simply polynomials. Compare the *degrees* (highest exponents) of $n(x)$ and $d(x)$:

- If the degree of the numerator is higher, then $\lim_{x\to\infty} r(x) = \infty$ or $-\infty$ (i.e., there is no limit because the function increases or decreases infinitely).

- If the degree of the denominator is higher, then $\lim_{x\to\infty} r(x) = 0$.

- If the degrees are equal, then $\lim_{x\to\infty} r(x)$ is equal to the *leading coefficient* of $n(x)$ divided by the leading coefficient of $d(x)$.

CRITICAL POINT

If $r(x)$ is a rational (fractional) function and has a horizontal asymptote, then it is guaranteed that $\lim\limits_{x \to \infty} r(x) = \lim\limits_{x \to -\infty} r(x)$. In other words, a rational function has the same horizontal asymptote at both ends of the function. However, once you include a radical, this guarantee goes down the drain.

Remember, these guidelines only apply to limits at infinity.

DEFINITION

The **degree** of a polynomial is the value of its largest exponent. The **leading coefficient** of a polynomial is the coefficient of the term with the largest exponent. For example, the expression $y = 3x^2 - 5x^6 + 7$ has degree 6 and leading coefficient –5.

Example 6: Evaluate $\lim\limits_{x \to \infty} \dfrac{5x^3 + 4x^2 - 7x + 4}{2 + x - 6x^2 + 8x^3}$.

Solution: This is a limit at infinity, so you should compare the degrees of the numerator and denominator. They are both 3, so the limit is equal to the leading coefficient of the numerator (5) divided by the leading coefficient of the denominator (8), so $\lim\limits_{x \to \infty} \dfrac{5x^3 + 4x^2 - 7x + 4}{2 + x - 6x^2 + 8x^3} = \dfrac{5}{8}$.

YOU'VE GOT PROBLEMS

Problem 5: Evaluate the following limits:

(a) $\lim\limits_{x \to \infty} \dfrac{2x^2 + 6}{3x^2 - 4x + 1}$

(b) $\lim\limits_{x \to -\infty} \dfrac{3x^2 + 4x + 3}{x^3 + 8x + 14}$

Now that you've got the basics down, let's turn the difficulty up a notch.

Example 7: Given the function $g(x)$ as defined here, assume that $\lim\limits_{x \to \infty} g(x) = \frac{1}{3}$. If $a + f = 12$, what is the value of $a - f$?

$$g(x) = \frac{ax^2 + bx + c}{d + ex + fx^2}$$

Solution: Remember, as with Example 5, all of the letters except for g and x (in other words a, b, c, d, e, and f) are just numbers.

Although there are letters all over the place in this fraction, you do have a few numbers to work with. Specifically, a couple of 2s, the exponents in the numerator and denominator. That means, as ugly as they are, the expressions in both parts of the fraction are just quadratics.

A fraction where the degrees of the numerator and denominator are equal? Sounds familiar! When those degrees are equal, the limit as x approaches infinity is equal to the leading coefficient of the numerator (a) divided by the leading coefficient of the denominator (f). The problem tells you that the limit is equal to $\frac{1}{3}$.

$$\lim_{x \to \infty} g\left(x\right) = \frac{a}{f} = \frac{1}{3}$$

Okay, that's a start, but you need more information to move forward. Luckily, the problem provides that information. It tells you that $a + f = 12$. Now you have two equations that explain the relationship between a and f; you have a system of equations that you can solve by substitution. Solve the linear equation for a by subtracting f from both sides:

$$a + f = 12$$
$$a = 12 - f$$

Because a has the same value as $12 - f$, you can replace a with $12 - f$ in the other equation, the proportion you created earlier:

$$\frac{a}{f} = \frac{1}{3}$$
$$\frac{12 - f}{f} = \frac{1}{3}$$

Finally! An equation with just one variable. Solve this for f by cross-multiplying.

$$3\left(12 - f\right) = 1 \cdot f$$
$$36 - 3f = f$$
$$36 = 4f$$
$$\frac{36}{4} = f$$
$$9 = f$$

Now that you know the value of f, you can find the corresponding value of a. Just use that equation you used to solve for a only a few moments ago:

$$a = 12 - f$$
$$= 12 - 9$$
$$= 3$$

That was a lot of work, but now we can finally answer the question posed by the problem. If $a + f = 12$, what is the value of $a - f$?

$$a - f = 3 - 9$$
$$= -6$$

> **YOU'VE GOT PROBLEMS**
>
> Problem 6: Identify a rational function $h(x)$ for which $\lim\limits_{x \to \infty} h(x) = 0$ and $\lim\limits_{x \to 3} h(x)$ do not exist.

Special Limit Theorems

The five following special limits are not special because of the way they make you feel all giddy inside. By "special," I mean they can't be evaluated by the means we've discussed so far, yet you'll see them frequently and should probably memorize them, even though that stinks. Now that we're on the same page, so to speak, here they are with no further ado:

- $\lim\limits_{\alpha \to 0} \frac{\sin \alpha}{\alpha} = 1$

This formula is only true when you approach 0, so don't use this under any other circumstances. The α can be any quantity.

- $\lim\limits_{\alpha \to 0} \frac{\cos \alpha - 1}{\alpha} = 0$

As with the first special limit, this formula is only true when approaching 0. Sometimes, you'll also see this formula written as $\frac{1 - \cos \alpha}{\alpha}$; the limit is still 0 either way.

- $\lim\limits_{x \to \infty} \frac{\text{any real number}}{x^{\text{any integer} > 0}} = 0$

If any real number is divided by x, and we let that x get infinitely large, the result is 0. Think about that—it makes sense. What is 4 divided by 900 kajillion? Who knows, but it's definitely very, very small. So small, in fact, that it's basically 0.

- $\lim\limits_{x \to \infty} \left(1 + \frac{1}{x}\right)^x = e$

This basically says that 1 plus an extremely small number, when raised to an extremely high power, is exactly equal to Euler's number (2.71828 …). You will see this very infrequently, but it's important to recognize it when you do.

- $\lim\limits_{\alpha \to 0} \frac{\tan \alpha}{\alpha} = 1$

This is another specialty trigonometric limit that pops up now and again. Tuck it into the folds of your brain in case it shows up on a test.

Example 8: Evaluate $\lim\limits_{x\to 0}\frac{\sin 3x}{x}$.

Solution: This is the first special limit formula, but notice that the value inside sine must match the denominator for that formula to work; therefore, we need a $3x$ in the denominator instead of just x. The trick is to multiply the top and bottom by 3 (because that's really the same thing as multiplying by 1—you're not changing the expression's value):

$$\lim_{x\to 0}\frac{\sin 3x}{x}\cdot\frac{3}{3}=\lim_{x\to 0}\frac{\sin 3x}{3x}\cdot 3$$

You can evaluate the limits of the factors separately and multiply the results together for the final answer:

$$\left(\lim_{x\to 0}\frac{\sin 3x}{3x}\right)\left(\lim_{x\to 0}3\right)=(1)(3)=3$$

YOU'VE GOT PROBLEMS

Problem 7: Evaluate $\lim\limits_{x\to\infty}\left(\frac{5}{x^3}+\left(1+\frac{1}{x}\right)^x\right)$.

Evaluating Limits Graphically

Throughout this chapter, you have focused on computational methods to calculate limits. Before you move on, it's time to reinforce everything you've learned by calculating limits graphically. Why do this? It may help you determine whether or not you understand what finite and infinite limits actually mean. Sometimes it's easy to fall into a "computational trap," where you get so focused on the step-by-step algorithms of calculus that you forget how the techniques connect to reality.

Example 9: Given the graph of $f(x)$ in Figure 6.3, evaluate the following limits:

(a) $\lim\limits_{x\to 0}f(x)$ (b) $\lim\limits_{x\to -\infty}f(x)$ (c) $\lim\limits_{x\to 1}f(x)$ (d) $\lim\limits_{x\to 2}f(x)$

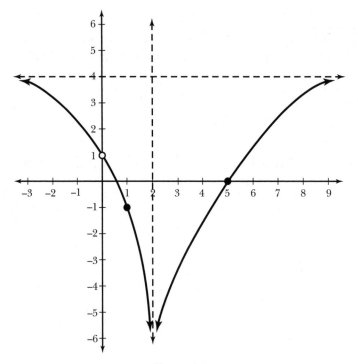

Figure 6.3

The graph of f(x). *Assume that the dotted lines are asymptotes.*

Solution:

(a) As you approach $x = 0$ from the right and the left, the graph approaches the hole at point $(0,1)$. Because the graph intends to reach a height of 1, $\lim_{x \to 0} f(x) = 1$.

(b) Trace your finger along the graph from right to left. As the graph heads off toward $-\infty$ at its left edge, it is trying to reach a height of 4. The horizontal asymptote at that height will prevent it from ever reaching $y = 4$, but that does not change the limit: $\lim_{x \to -\infty} f(x) = 4$.

(c) As you approach $x = 1$ from the left and the right, the graph intends to reach (and actually *does* reach) a height of –1. Thus, $\lim_{x \to 1} f(x) = -1$.

(d) Whether you approach $x = 2$ from the left or the right, the graph plummets, decreasing without bound. Therefore, the limit does not exist, because $\lim_{x \to 2} f(x) = -\infty$.

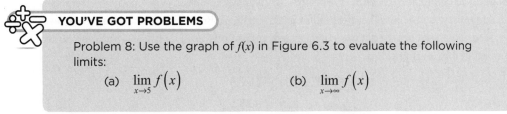

YOU'VE GOT PROBLEMS

Problem 8: Use the graph of $f(x)$ in Figure 6.3 to evaluate the following limits:

(a) $\lim\limits_{x \to 5} f\left(x\right)$

(b) $\lim\limits_{x \to \infty} f\left(x\right)$

Technology Focus: Calculating Limits

Most calculus courses these days allow you to use a graphing calculator or other technology tool, at least some of the time. The tools vary widely in their functionality, but the process we'll explore here works the same for all calculators (except for a few tricks I included for Texas Instruments calculator owners).

How can you use graphing calculators to check your homework on limit problems? Well, calculators don't mind plugging really small, annoying numbers into expressions—and we love them for their embrace of the mundane.

Back in Example 3, you applied the conjugate method to calculate a fairly complicated limit:

$$\lim_{x \to 5} \frac{\sqrt{x+11}-4}{x-5} = \frac{1}{8}$$

The problem asks you to calculate a limit as x approaches 5, so plug an x-value *very close* to 5 into the fraction, such as $x = 5.0001$:

$$\frac{\sqrt{5.0001+11}-4}{5.0001-5} = \frac{\sqrt{16.0001}-4}{0.0001}$$

I don't know about you, but I can't calculate $\sqrt{16.0001}$ off the top of my head, let alone subtract 4 from it and then divide it by 0.0001. Your calculator, however, is more than happy to do the work for you:

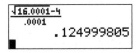

Figure 6.4
Your calculator may be able to express mathematical expressions that look like this, or your screen might look more like Figure 6.5.

Your digital buddy reports an approximate limit of 0.124999805, which is very close to 0.125, or $\frac{1}{8}$.

By the way, the calculator screen in Figure 6.4 formats mathematics very neatly. The fractions and the square roots look just like they do in a regular math book. However, your calculator doesn't need to be so fancy to get the correct answer. You just need to use parentheses carefully, as demonstrated in Figure 6.5.

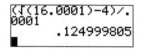

Figure 6.5

This screen isn't formatted as neatly, but you still get the same answer.

So, your calculator can help you check answers for problems that approach a finite limit, in this case $x = 5$. What about limits at infinity?

In Example 6, you calculated a limit as x approaches infinity of a rational function:

$$\lim_{x \to \infty} \frac{5x^3 + 4x^2 - 7x + 4}{2 + x - 6x^2 + 8x^3} = \frac{5}{8}$$

Near that problem, in a "Critical Point" sidebar, I insisted that if a rational function has a limit as x approaches ∞, then it has the same limit as x approaches $-\infty$. Let's use your calculator to make sure I'm not lying to you and verify that the limit as x approaches $-\infty$ is the same: $\frac{5}{8}$.

Replace x in the rational function with a really large negative number, something like $x = -10,000$. The result should come back close to 0.625, the decimal equivalent of $\frac{5}{8}$:

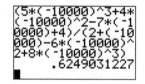

Figure 6.6

There are a lot of zeros on this screen, so it's pretty confusing. If only there were a better way ....

The final result, .6249031227, proves that I'm not a liar (at least not about asymptotes). The limit as x approaches $-\infty$ is $\frac{5}{8}$. Let's be honest, though. That was a lot of typing, which leaves a lot of places to make mistakes. Luckily, there is a shortcut!

Your calculator probably lets you store values of variables in memory. On the Texas Instruments family of calculators, the STO▸ button tells the calculator to store a number as a variable. For example, you could set $A = -10,000$ by typing "10000 STO▸ ALPHA MATH." (Pressing the green ALPHA button allows you to access the green letters above the buttons, and the green letter A appears above the MATH button.) Your screen should look like Figure 6.7. If you own a different calculator, check the manual for more information and instructions.

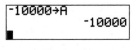

Figure 6.7
This set of keystrokes sets the variable A *equal to –10,000 in your calculator's memory.*
From now on, whenever it sees "A," it thinks "–10,000."

Now that you have stored –10,000 as *A*, you can type the rational expression using *A*, which is much easier on the eyes; the result is the same.

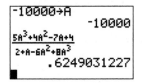

Figure 6.8
This is the same calculation as Figure 6.5, but this time the value –10,000 is represented by A.

The Least You Need to Know

- Most limits can be evaluated via the substitution, factoring, or conjugate methods.

- If a function $f(x)$ has a vertical asymptote $x = c$, then $\lim\limits_{x \to c} f(x) = \infty$ or $-\infty$.

- If a rational function $f(x)$ has a horizontal asymptote $y = L$, then $\lim\limits_{x \to \infty} f(x) = \lim\limits_{x \to -\infty} f(x) = L$ (as long as there are no radicals in there).

- There are five common limits that defy our techniques and must be memorized.

CHAPTER 7

Continuity

Now that you understand and can evaluate limits, it's time to move forward with that knowledge. Flip through any calculus textbook and read some of the most important calculus theorems, and you'll find that nearly every one contains a significant condition: continuity. In fact, almost none of our most important calculus conclusions (including the Fundamental Theorem of Calculus, which sounds pretty darn important) work if the functions in question are not continuous.

Testing for continuity on a function is very similar to testing for the existence of limits on a function. Just as three stipulations must be met in order for a limit to exist at a given point (left- and right-hand limits existing and being equal), three different stipulations must be met in order for a function to be continuous at a point. Just as there are three major cases in which limits do not exist, there are three major causes that force a function to be discontinuous.

Calculus is handy like that—if you look hard enough, you can usually see how one topic flows into the next. Without limits, there'd be no continuity; without continuity, there'd be no derivatives; without derivatives, no integrals; and without integrals, no sleepless, panicked nights trying to cram for calculus tests.

In This Chapter

- What it means to be continuous
- Classifying discontinuity
- When discontinuity is removable
- The Intermediate Value Theorem

What Does Continuity Look Like?

First of all, let's set our language straight. *Continuous* is an adjective that describes a function meeting very specific standards. Just as Boy Scouts must pass tests to earn merit badges, there are three tests a function must pass at any given point in order to earn the "Continuous" merit badge.

Before we get into the nitty-gritty of the math definition, let's approach continuity from a visual perspective. It is easiest to determine whether or not a function is continuous by looking at its graph. *If the graph has no holes, breaks, or jumps, then we can rest assured that the function is continuous.* A continuous function is simply a nice, smooth function that can be drawn completely without lifting your pencil. With this intuitive definition in mind, study Figure 7.1. Can you tell which of the following three functions is continuous?

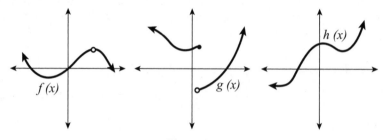

Figure 7.1

One of these things is not like the others; one of these things just doesn't belong. Which is the continuous function?

✏️ **CRITICAL POINT**

A continuous function is like a well-built roller coaster track—no gaps, holes, or breaks means safe riding for its passengers.

Of these functions, only $h(x)$ can be drawn with a single, unbroken pen stroke. The other functions are much more unpredictable: f has an unexpected hole in it, and g suddenly breaks without warning. Only good old h provides a nice smooth ride from start to finish. Function h is like the good, solid, dependable boyfriend or girlfriend you wouldn't be embarrassed to bring home to Mom and Dad, and the fact that it guarantees no unexpected breakups means peace of mind for your emotional well-being.

The Mathematical Definition of Continuity

The mathematical definition of continuity makes a lot of sense if you keep one thing in mind: whereas limits told us where a function intended to go, continuity guarantees that the function actually made it there. As the saying goes, "The road to hell is paved with good intentions."

Continuity has the mathematical role of policeman, determining whether or not the function followed through with its intentions (meaning it is continuous) or not (making the function discontinuous).

Many functions are guaranteed to be continuous at each point in their domain, including polynomial, radical, exponential, logarithmic, rational, and trigonometric functions. Most of the discontinuous functions you'll encounter will be due to undefined spots in rational functions and jumps due to piecewise-defined functions. We'll discuss more about specific causes of discontinuity in the next section.

> **CRITICAL POINT**
>
> A function is continuous at a point if the limit and function value there are equal. In other words, the limit exists if the intended height matches the actual function height.

With that in mind, here is the official definition of *continuity*:

A function $f(x)$ is continuous at a point $x = c$ if the following three conditions are met:

- $\lim\limits_{x \to c} f\left(x\right)$ exists

- $f(c)$ is defined

- $\lim\limits_{x \to c} f\left(x\right) = f\left(c\right)$

In other words, the limit exists at $x = c$ (which means the function has an intended height); the function exists at $x = c$ (which means that there is no hole there); and the limit is equal to the function value (i.e., the function's value matches its intended value). (By the way, if a function is continuous, you can evaluate any limit on that function using the substitution method, since the function's value at any point will be equal to the limit there.)

Example 1: Explain why the function $f(x)$, defined here, is continuous at $x = -3$.

$$f\left(x\right) = \begin{cases} \frac{\sqrt{x+19}-4}{x+3}, & x \neq -3 \\ \frac{1}{8}, & x = -3 \end{cases}$$

Solution: To test for continuity, you must find the limit and the function value at $x = -3$ (and make sure they are equal). Now, that's one ugly function. How can you determine its intended height (limit) at $x = -3$? Clearly, the top rule in this piecewise-defined function governs the function's value for every single x except for $x = -3$. When you are finding a limit, you want to see what height is intended as you approach $x = -3$, not the value actually reached at $x = -3$, so you'll find the limit of the larger, ugly, top rule for f. Use the conjugate method:

$$\lim_{x \to -3} \frac{\sqrt{x+19}-4}{x+3} \cdot \frac{\sqrt{x+19}+4}{\sqrt{x+19}+4}$$

$$= \lim_{x \to -3} \frac{\left(\sqrt{x+19}\right)^2 + 4\sqrt{x+19} - 4\sqrt{x+19} - 4^2}{(x+3)\left(\sqrt{x+19}+4\right)}$$

$$= \lim_{x \to -3} \frac{x+19-16}{(x+3)\left(\sqrt{x+19}+4\right)}$$

$$= \lim_{x \to -3} \frac{\cancel{x+3}}{\cancel{(x+3)}\left(\sqrt{x+19}+4\right)}$$

$$= \lim_{x \to -3} \frac{1}{\sqrt{x+19}+4}$$

Whew! Now that $(x + 3)$ no longer appears in the denominator, you can apply the substitution method:

$$= \frac{1}{\sqrt{-3+19}+4}$$

$$= \frac{1}{\sqrt{16}+4}$$

$$= \frac{1}{8}$$

The limit clearly exists when $x = -3$, and it is equal to $\frac{1}{8}$. The first condition of continuity is satisfied. Now, on to the second. According to the function's definition, you know that $f(-3) = \frac{1}{8}$, so the function does exist there. Therefore, you can conclude that the function is continuous at $x = -3$, because the limit is equal to the function value there.

Example 2: Determine whether the function $g(x)$, defined here, is continuous at $x = \frac{1}{2}$:

$$g(x) = \begin{cases} \frac{2x^2+3x-2}{2x-1}, & x < \frac{1}{2} \\ \frac{8x-1}{2}, & x \geq \frac{1}{2} \end{cases}$$

Solution: If $g(x)$ is continuous at $x = \frac{1}{2}$, three things have to be true: (1) $\lim_{x \to 1/2} g(x)$ must exist, (2) $g\left(\frac{1}{2}\right)$ must exist, and (3) those values must be equal. The easiest of the three conditions to test is the second one. Calculate $g\left(\frac{1}{2}\right)$:

$$g\left(\frac{1}{2}\right) = \frac{8(1/2)-1}{2}$$

$$= \frac{4-1}{2}$$

$$= \frac{3}{2}$$

So far so good; $g\left(\frac{1}{2}\right)$ exists, but now there's no avoiding the much more difficult task of determining whether the limit exists there. Notice that the way $g(x)$ is defined changes at $x = \frac{1}{2}$, which means $g(x)$ may intend to reach two different heights on the graph as you approach from the left and the right. Good news! You have already calculated the right-hand limit using the substitution method:

$$\lim_{x \to 1/2^+} g(x) = \lim_{x \to 1/2} \frac{8x-1}{2}$$
$$= \frac{3}{2}$$

Evaluating the left-hand limit is a little trickier. You need to employ the factoring method:

$$\lim_{x \to 1/2^-} g(x) = \lim_{x \to 1/2} \frac{2x^2+3x-2}{2x-1}$$
$$= \lim_{x \to 1/2} \frac{(2x-1)(x+2)}{2x-1}$$
$$= \lim_{x \to 1/2} \frac{\cancel{(2x-1)}(x+2)}{\cancel{2x-1}}$$
$$= \lim_{x \to 1/2} (x+2)$$
$$= \frac{1}{2} + 2$$
$$= \frac{1}{2} + \frac{4}{2}$$
$$= \frac{5}{2}$$

Red alert! The left- and right-hand limits are not equal as x approaches $\frac{1}{2}$, because $\frac{3}{2} \neq \frac{5}{2}$. If the left- and right-hand limits aren't equal, then the general limit doesn't exist at $x = \frac{1}{2}$. That means the first condition of continuity has not been met.

Your goal was to determine whether or not $g(x)$ is continuous when $x = \frac{1}{2}$, and the answer is no. Take a look at the graph of $g(x)$ to visually verify that the function does not intend to reach the same location as you approach $x = \frac{1}{2}$ from the left and the right.

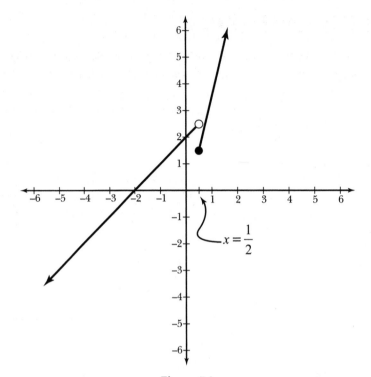

Figure 7.2

Function g(x) is not continuous at $x = \frac{1}{2}$. The graph also helps you decide which expression to use as you approach from the left and the right: the graph of $\frac{2x^2+3x-2}{2x-1}$ is left of $x = \frac{1}{2}$ and the graph of $\frac{8x-1}{2}$ is right of $x = \frac{1}{2}$.

A function cannot be continuous at an *x*-value where a limit does not exist; *g(x)* is discontinuous at $x = \frac{1}{2}$ due to a *jump discontinuity*. Are you asking yourself, "What's a jump discontinuity?" If so, I'm glad you asked. If not, play along and pretend you did. Either way, I'll meet you in the next section to explain what I mean.

YOU'VE GOT PROBLEMS

Problem 1: Determine whether or not the function *g(x)*, defined here, is continuous at *x* = 1.

$$g(x) = \begin{cases} \frac{3x^2-x-2}{x-1}, & x \neq 1 \\ -2, & x = 1 \end{cases}$$

Types of Discontinuity

Not much happens in the life of a graph—it lives in a happy little domain, playing matchmaker to pairs of coordinates. However, there are three things that can happen over the span of a function which change it fundamentally, making it discontinuous. Memorizing the three major causes of discontinuity is not so important; instead, recognize exactly what causes the function to fall short of continuity's requirements.

Jump Discontinuity

A *jump discontinuity* is typically caused by a piecewise-defined function whose pieces don't meet neatly, leaving gaping tears in the graph large enough for an elephant, or other tusked mammal, to walk through. Consider the function:

$$f(x) = \begin{cases} -x+3, & x<0 \\ x+1, & x\geq0 \end{cases}$$

This graph is made up of two linear pieces, and the rule governing the function changes when $x = 0$. Look at the graph of $f(x)$ in Figure 7.3.

 **DEFINITION**

A **jump discontinuity** occurs when no general limit exists at the given x value (because the right- and left-hand limits exist but are not equal). A function is *everywhere continuous* if it is continuous for each x in its domain.

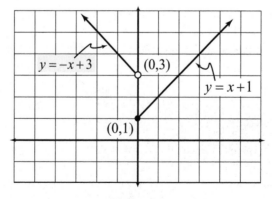

Figure 7.3
The graph of f(x) *exhibits an unhealthy dual personality since it is defined by a piecewise-defined function.*

When $x = 0$, the graph has a tragic and unsightly break. Whereas the left-hand piece is heading toward a height of $y = 3$ as you approach $x = 0$, the right-hand piece has a height of $y = 1$ when $x = 0$. Does this sound familiar? It should: the left- and right-hand limits are unequal at $x = 0$, so $\lim_{x \to 0} f(x)$ does not exist. This breaks the first rule requirement of continuity, rendering $f(x)$ discontinuous.

In the next example, you are given a piecewise-defined function. Your goal is to shield it from the same fate as the pitiful function $f(x)$ by choosing a value for the constant c that ensures that the pieces of the graph will meet when the defining function rule changes. Godspeed!

Example 3: Given $h(x)$ as defined here, calculate the real number c that makes $h(x)$ everywhere continuous:

$$h(x) = \begin{cases} x - 2, & x \leq 3 \\ x^2 + c, & x > 3 \end{cases}$$

Solution: The top rule in $h(x)$ will define the function for all numbers less than or equal to 3, and its reign ends once x reaches that boundary. At that point, $h(x)$ will have reached a height of $h(3) = 3 - 2 = 1$. Therefore, the next rule ($x^2 + c$) must start at *exactly* that height when $x = 3$, even though it is technically defined only when $x > 3$. That's the key: both pieces must reach the exact same intended height when the graph of a piecewise-defined function changes rules. Thus, when $x = 3$:

$$x^2 + c = 1$$

Plug in that x value and solve for c:

$$3^2 + c = 1$$
$$9 + c = 1$$
$$c = -8$$

Thus, the second piece of $g(x)$ must be $x^2 - 8$ in order for $h(x)$ to be continuous. You can verify the solution with the graph of $h(x)$ (as shown in Figure 7.4)—no jump discontinuity anywhere to be found.

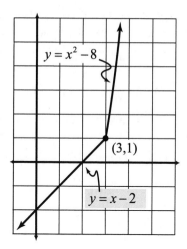

$y = x^2 - 8$

(3,1)

$y = x - 2$

Figure 7.4
The graph of h(x) *is nice and continuous now—the pieces of the graph join together seamlessly.*

Example 4: Given the function $q(x)$ defined here, calculate the values of a and b that ensure $q(x)$ is continuous for all real numbers. Verify your answers with a graph of $q(x)$.

$$q(x) = \begin{cases} -x^2, & x \le -2 \\ a, & -2 < x < 2 \\ ax + b, & x \ge 2 \end{cases}$$

Solution: The individual pieces of function $q(x)$ are each continuous over their domains. The graphs of $y = -x^2$ (a parabola), $y = a$ (a horizontal line at height a), and $y = ax + b$ (a line with slope a and y-intercept b) are unbroken and free of jump discontinuities. Of course, this is a piecewise-defined function, so each of the pieces need to meet when $x = -2$ and $x = 2$ if the overall function is going to be continuous. You need to pay special attention to these x-values, where the rules that define the function change.

Let's start with $x = -2$. If you substitute that value into $-x^2$, you calculate the left-hand limit of $q(x)$ as x approaches -2.

$$\lim_{x \to -2^-} q(x) = -(-2)^2$$
$$= -(4)$$
$$= -4$$

The right-hand limit at $x = -2$ *must* be equal to the left-hand limit for the general limit to exist. The right-hand limit is simply the constant a. If the right- and left-hand limits are equal, you can conclude that a also equals -4.

Not only is $a = -4$ the limit as you approach $x = -2$ from the right, it is also the limit as you approach $x = 2$ from the left:

$$\lim_{x \to 2^-} q(x) = -4$$

In order for a limit to exist at $x = 2$, the right-hand limit must also equal -4:

$$\lim_{x \to 2^+} q(x) = -4$$

That right-hand limit is equal to the expression $ax + b$, when $x = 2$:

$$a(2) + b = -4$$

Recall that $a = -4$ and solve for b:

$$-4\left(2\right) + b = -4$$
$$-8 + b = -4$$
$$b = 8 - 4$$
$$b = 4$$

Consider the graph of $q(x)$ in Figure 7.5. Notice that there are no jump discontinuities at $x = -2$ or $x = 2$.

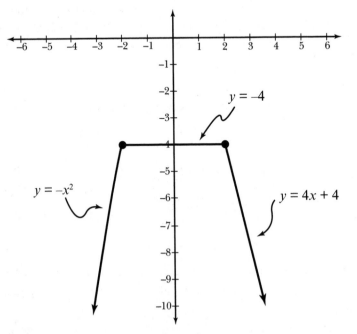

Figure 7.5
The graph of q(x) *is continuous for all real numbers, including* x $= -2$ *and* x $= 2$.

YOU'VE GOT PROBLEMS

Problem 2: Calculate the value of a that makes the function $h(x)$ everywhere continuous, given:

$$h(x) = \begin{cases} 2x^2 + x - 7, & x < -1 \\ ax + 6, & x \geq -1 \end{cases}$$

Point Discontinuity

Point discontinuity occurs when a function contains a hole. Think of it this way: the function is discontinuous only because of that rascally little point, hence the name.

DEFINITION

A **point discontinuity** occurs when a general limit exists, but the function value is not defined there, breaking the second condition of continuity.

Consider the function $p(x) = \frac{x^2 + 11x + 28}{x + 4}$. It is a rational function, so it's continuous for all points in its domain. Hold on a second, though. The value $x = -4$ is definitely not in the domain of $p(x)$ (look at the denominator), so $p(x)$ will automatically be discontinuous there. The question is: what sort of discontinuity is present?

Classifying the discontinuity in this case is very easy—all you have to do is to test for a limit at that x value. To calculate the limit, use the factoring method:

$$\lim_{x \to -4} \frac{x^2 + 11x + 28}{x + 4}$$
$$= \lim_{x \to -4} \frac{(x + 4)(x + 7)}{x + 4}$$
$$= \lim_{x \to -4} (x + 7)$$
$$= -4 + 7$$
$$= 3$$

In conclusion, because $x = -4$ represents a place where $p(x)$ is undefined, and $\lim_{x \to -4} p(x) = 3$, you know that there is a hole in the function $p(x)$ at the point $(-4, 3)$, a point discontinuity. See the graph of $p(x)$ in Figure 7.6.

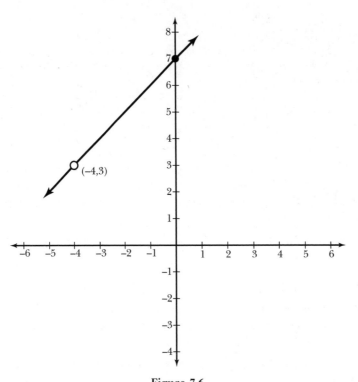

Figure 7.6
The graph of p(x) *is also the graph of* y = x + 7, *except when* x = −4, *where the function is undefined.*

✏️ **CRITICAL POINT**

Any *x*-value for which a function is undefined will automatically be a point of discontinuity for the function. If a limit exists at the point of discontinuity, then it must be a point discontinuity.

Infinite/Essential Discontinuity

An *infinite* (or *essential*) *discontinuity* occurs when a function neither has a limit (because the function increases or decreases without bound) nor is defined at the given *x*-value. In other words, this type of discontinuity occurs primarily at a vertical asymptote.

DEFINITION

An **infinite discontinuity** is caused by a vertical asymptote. Since the function increases or decreases without bound, there can be no limit, and since the function never actually touches the asymptote, the function is undefined there. Thus, the presence of a vertical asymptote ruins all the conditions necessary for continuity to occur. Vertical asymptotes are the home wreckers of the continuity world.

It's easy to determine which x-values cause a vertical asymptote if you remember the shortcut from the last chapter: a function increases or decreases infinitely at a given value of x if substituting that x into the expression results in a constant divided by 0. On the other hand, a result of $\frac{0}{0}$ *usually* means that point discontinuity is at work. However, since a result of $\frac{0}{0}$ doesn't *guarantee* you've got point discontinuity, you'll need to double-check to see if the limit exists there. We'll do this in Example 5, in case you're confused.

In summary: if no general limit exists, you have jump discontinuity; if the limit exists but the function doesn't, you have point discontinuity; if the limit doesn't exist because it is ∞ or $-\infty$, you have infinite discontinuity.

Now that you've got the field guide to discontinuity, let's look at a typical problem you'll be given. In it, you'll be asked either to identify where a function is continuous or, instead, to highlight areas of discontinuity and to classify the type of discontinuity present.

Example 5: Give all x-values for which the function $f(x)$, defined here, is discontinuous, and classify each instance of discontinuity.

$$f\left(x\right) = \frac{9x^2 - 3x - 2}{3x^2 + 13x + 4}$$

Solution: This is a rational function, so it's guaranteed to be continuous on its entire domain; the only points you have to inspect are where $f(x)$ is undefined. Because $f(x)$ is rational, it is undefined when its denominator equals 0, and the easiest way to find those locations is by factoring $f(x)$:

$$f\left(x\right) = \frac{\left(3x+1\right)\left(3x-2\right)}{\left(x+4\right)\left(3x+1\right)}$$

Set the denominator equal to 0 to see that $x = -4$ and $x = -\frac{1}{3}$ will be points of discontinuity. Now, we need to explain what kinds of discontinuity they represent. Plug each into $f(x)$. Substituting $x = -4$ results in $\frac{154}{0}$, indicating the presence of a vertical asymptote and an infinite discontinuity. However, substituting $x = -\frac{1}{3}$ into $f(x)$ gives you $\frac{0}{0}$, which means there is probably a hole in the function there. That's not good enough supporting work, however. You need to *prove* that there's a hole there in order to conclude that $x = -\frac{1}{3}$ represents a point discontinuity. All the proof you'll need is to verify the presence of a limit at $x = -\frac{1}{3}$. To do so, use the factoring method:

$$\lim_{x \to -1/3} \frac{(3x+1)(3x-2)}{(x+4)(3x+1)}$$

$$= \lim_{x \to -1/3} \frac{3x-2}{x+4}$$

$$= \frac{3(-1/3)-2}{-1/3+4}$$

$$= \frac{-1-2}{-1/3+12/3}$$

$$= \frac{-3}{11/3}$$

Convert the complex fraction into a regular fraction by multiplying the numerator (−3) by the reciprocal of the denominator $\left(\frac{3}{11}\right)$:

$$= -3 \cdot \frac{3}{11}$$

$$= -\frac{9}{11}$$

Because the limit exists, there is a point discontinuity when $x = -\frac{1}{3}$. You can verify this with the graph of $f(x)$ in Figure 7.7.

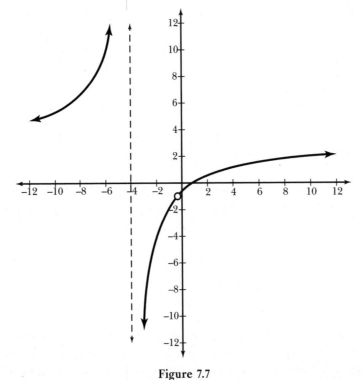

Figure 7.7

The graph of f(x) *has an infinite discontinuity at* x = −4 *and a point discontinuity at* $x = -\frac{1}{3}$.

YOU'VE GOT PROBLEMS

Problem 3: Give all x-values for which the function here is discontinuous, and classify each instance of discontinuity.

$$g\left(x\right) = \frac{2x^2 + 5x - 25}{x^2 - 25}$$

Removable vs. Nonremovable Discontinuity

Occasionally you'll see a function described as having removable or nonremovable discontinuity. These terms are more specific than simply stating that a function is discontinuous, but less specific than stating that it has point, jump, or infinite discontinuity.

However, because these terms appear often, it's good to know what they mean. A *removable discontinuity* is one that could be eliminated by simply redefining a finite number of points. In other words, if you can "fix" the discontinuity by filling in holes, then that discontinuity is removable. Let's go back to the function we examined in Example 5 for a moment:

$$f\left(x\right) = \frac{(3x+1)(3x-2)}{(x+4)(3x+1)}$$

 **DEFINITION**

A **removable discontinuity** occurs at a given x-value if a limit exists there, since you can redefine the function to fill in the holes (and thus *remove* the discontinuity) if you choose to. Therefore, point and removable discontinuity are essentially synonymous.

This function has a point discontinuity at $x = -\frac{1}{3}$, because $\lim\limits_{x \to -1/3} f\left(x\right) = -\frac{9}{11}$. If I redefine $f(x)$ slightly, setting $f\left(-\frac{1}{3}\right) = -\frac{9}{11}$, then the limit equals the function value when $x = -\frac{1}{3}$, and $f(x)$ is continuous there. Mathematically, the new function $f(x)$ looks like this:

$$f\left(x\right) = \begin{cases} \frac{(3x+1)(3x-2)}{(x+4)(3x+1)}, & x \neq -\frac{1}{3} \\ -\frac{9}{11}, & x = -\frac{1}{3} \end{cases}$$

You don't actually have to change the function for it to be removably discontinuous (in fact, if you did change the function, it wouldn't be discontinuous when $x = -\frac{1}{3}$). However, if it is *possible* to change a few points in order to fill in the function's holes, the discontinuities are removable.

Nonremovable discontinuity occurs when a function has no general limit at the given *x*-value, as is the case with infinite and jump discontinuities. There is no way to redefine a finite number of points to "repair" this type of discontinuity; the function is fundamentally discontinuous there, and no amount of rehabilitation or mood-altering medication can make this function safe for a cultured society. I gasp at the thought, but not even a charity rock concert can help (sorry, U2). Back to the function *f*(*x*) from Example 5 for illustration. Since a vertical asymptote occurs at *x* = −4 and no general limit exists there, *x* = −4 represents an instance of nonremovable discontinuity.

> **DEFINITION**
>
> A **nonremovable discontinuity** occurs at a given *x*-value if no general limit exists there, making it impossible to remove the discontinuity by redefining a fixed number of points. Jump and infinite discontinuities are both examples of nonremovable discontinuity.

The Intermediate Value Theorem

Break out the party favors—we've arrived at our first official calculus theorem.

The Intermediate Value Theorem: If a function *f*(*x*) is continuous on the closed interval [*a*,*b*], then for every real number *d* between *f*(*a*) and *f*(*b*), there exists a real number *c* between *a* and *b* such that *f*(*c*) = *d*.

Now, let me explain what the heck that means using a simple example. Like all red-blooded Americans, I enjoy a little too much holiday dining during the winter months of November and December. If we were to exaggerate my weight gain (a little), it might have the following humorously titled "Kelley's Date vs. Weight Graph," which I'll call *w*(*x*) in Figure 7.8.

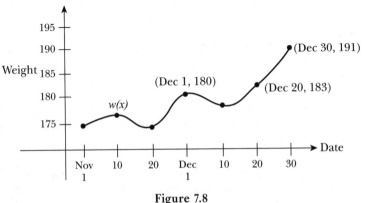

Figure 7.8

Kelley's Date vs. Weight Graph.

From the graph, we can see that I weighed 180 pounds on December 1 and porked up to 191 by the time December 30 "rolled around," poor pun definitely intended. Comparing this to the Intermediate Value Theorem, a = Dec 1, b = Dec 30 (weird values, but go with me on this), $f(a) = w(\text{Dec 1}) = 180$, and $f(b) = w(\text{Dec 30}) = 191$. According to the theorem, I can choose any value between 180 and 191 (for example, 183), and I am guaranteed that at some time between December 1 and December 30, I actually weighed that much.

The Intermediate Value Theorem does not claim to tell you *where* your function reaches that value or *how many times* it does. The theorem simply claims (in a calm, soothing voice) that every height a function reaches on a specific x-interval boundary will be output at least once by some x within that interval. As it only guarantees the existence of something, it is called an *existence theorem*.

Example 6: Explain why the function $j(x) = x^4 + 3x^2 - x - 6$ *must* have a root (x-intercept) on the interval $[-2,-1]$.

Solution: To get more information about the interval, substitute its endpoints into $j(x)$:

$$j(-2) = (-2)^4 + 3(-2)^2 - (-2) - 6 \qquad j(-1) = (-1)^4 + 3(-1)^2 - (-1) - 6$$
$$= 16 + 12 + 2 - 6 \qquad\qquad\qquad = 1 + 3 + 1 - 6$$
$$= 24 \qquad\qquad\qquad\qquad\qquad = -1$$

Function $j(x)$ is continuous for its entire domain—there are no denominators that might equal zero and cause a vertical asymptote, no radical expressions that might be negative under a square root sign and become undefined, no piecewise-defined functions that might cause a break in the graph. All you have here is a nice, smooth, unbroken graph of a polynomial, and as long as the function is continuous over a closed domain, like $[-2,-1]$, you can apply the Intermediate Value Theorem.

Watch carefully now. This problem is a lot like a magic trick—if you blink you might miss it. Here we go.

You know that the function $j(x)$ passes through point $(-2,24)$ and $(-1,-1)$. According to the Intermediate Value Theorem, every y-value between 24 and -1 corresponds to some x-value between -2 and -1. That's it—you're done!

Did you miss it? Every y-value between 24 and -1, *including 0*, has a corresponding x-value between -2 and -1. In other words, at some x-value on the interval $[-2,-1]$, y has a value of 0, causing a root or x-intercept in the graph.

Don't get all muddled up in the semantics of the theorem. Here's the gist of it. Imagine that there's a huge line on the floor of the room you're currently in, a line that divides the room roughly in half. Well, if at 10 A.M. you're standing on one side of that line and at 11 A.M. you're standing on the other side, then at some point between 10 and 11 A.M. (assuming you didn't leave the room or teleport through space and time) you crossed the line.

If you are particularly interested, this happens at approximately $x = -1.08$, as you can see in the graph of $j(x)$ in Figure 7.9.

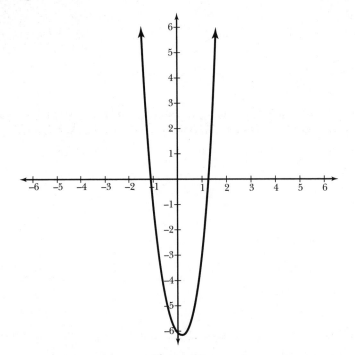

Figure 7.9
Function j(x) *also has a root between* x *= 1 and* x *= 2.*

YOU'VE GOT PROBLEMS

Problem 4: Use the Intermediate Value Theorem to explain why the function $g(x) = x^2 + 3x - 6$ *must* have a root on the interval [1,2].

The Least You Need to Know

- A continuous function has no holes, jumps, or breaks in its graph.
- If a function reaches its intended height at a particular x-value, the function is continuous there.
- If a function is undefined but possesses a limit at a given x-value, there is point discontinuity, which is removable.
- Infinite discontinuity is caused by a vertical asymptote, whereas jump discontinuity is caused by a break in the function's graph; both are nonremovable discontinuities.
- The Intermediate Value Theorem uses rather complex language to guarantee the "wholeness" or "completeness" of a graph.

CHAPTER 8

The Difference Quotient

Although limits are important to the development of calculus and are the only topic we have even discussed so far, they are about to take a backseat to the two major topics comprising what most people call "calculus": derivatives and integrals. It would be rude (and actually mathematically inaccurate) to simply start talking about derivatives without describing their relationship to limits.

Brace yourself. This chapter describes the solution to one of the most puzzling mathematical dilemmas of all time: how to calculate the slope of a tangent line to a nonlinear function. We're going to use limits to concoct a general formula that will allow you to find the tangent slope to a function at any given point. The process is a little tedious and is a bit algebra-intensive. You may ask yourself, "Am I always going to go through so much pain to find a derivative?" The answer is no. In Chapter 9, you'll learn lots of shortcuts to finding derivatives.

For now, however, prepare to be dazzled. You're about to create a tangent line to a function and calculate its slope through a little "mathemagics."

In This Chapter

- Creating a tangent from scratch
- How limits can calculate slope
- "Secant you shall find" the tangent line
- Both versions of the difference quotient

When a Secant Becomes a Tangent

Before we go about calculating the slope of a tangent line, you should probably know what a tangent line is. A *tangent line* is a line that just barely skims across the edge of a curve, hitting it at the point you intend.

> **DEFINITION**
>
> A **tangent line** skims across the curve, hitting it once in the indicated location; however, a secant line does not skim at all. It cuts right through a function, usually intersecting it in multiple spots.

In Figure 8.1, you'll see the graph of $y = \sin x$ with two of its tangent lines drawn, one at $x = \frac{\pi}{2}$ and one at $x = \frac{7\pi}{4}$. Notice that the tangent lines barely skim across the edge of the graph and hit only at one point, called the *point of tangency*. If you extend it, the tangent line may hit the function again somewhere else along the graph, but that doesn't matter. What matters is that it only hits once relatively close to the point of tangency.

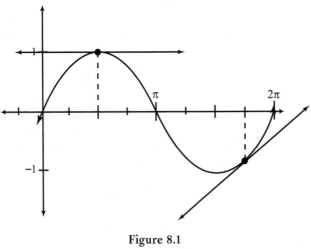

Figure 8.1
Points of tangency.

A *secant line*, on the other hand, is a line that crudely hacks right through a curve, usually hitting it in at least two places. In Figure 8.2, I have drawn both a secant and a tangent line to a function $f(x)$ when $x = 3$. Notice that the dotted secant line doesn't have the finesse of the tangent line, which strikes only at $x = 3$.

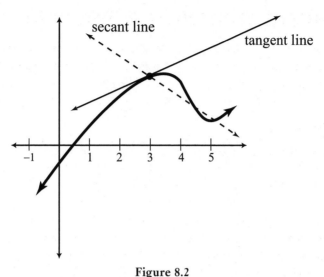

Figure 8.2
A secant and a tangent line to a function f(x) *when* x = 3.

Through a little trickery, we are going to make a secant line into a tangent. This is the backbone of our procedure for calculating the slope of a tangent line. So now that you know what the words mean, let's get started.

Honey, I Shrunk the Δx

Take a look at the function graph in Figure 8.3 called $f(x)$. I have marked the location $x = c$ on the graph. My final, overall goal will be to calculate the slope of the tangent line to f at $x = c$. You may not understand this to be a very important goal, but trust me, it is world-shatteringly important.

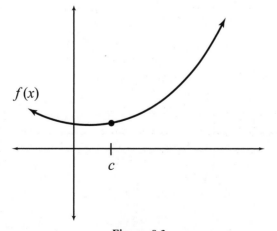

Figure 8.3
The graph of some function f(x) *with location* x = c.

Now, let's add a few things to the graph to create Figure 8.4. First of all, I know the coordinates of the indicated point. In order to get to that point from the origin, I have to go c units to the right and $f(c)$ units up (so that I hit the function), which translates to the coordinate pair $(c, f(c))$. Now let's add another point to the graph to the right of the point at $x = c$. How far to the right, you ask? Let's be generic and call it "Δx" more to the right. ("Δx" is math language for "the change in x," and since we're changing the x value of c by going Δx more to the right, it's a fitting name.)

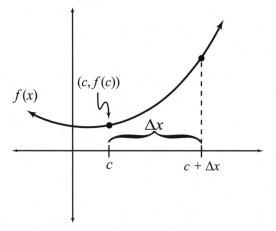

Figure 8.4

Now appearing on the graph of f(x), *a new* x *value, which is a distance of* Δx *away from* c.

Once again, all we're doing is making a new point that is a horizontal distance of Δx away from the first point. Can you figure out the coordinates of the new point? In the same fashion that we got the first coordinate pair to be $(c, f(c))$, this point has coordinates $(c + \Delta x, f(c + \Delta x))$. Now connect these two points together, and what have you got? A secant line through f, as pictured in Figure 8.5.

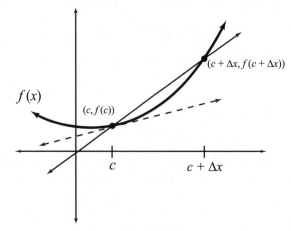

Figure 8.5

We've added the coordinates to the new point. Note that the secant line connecting the points looks a little like the dotted tangent line at x = c. *It's a little too steep, but it's pretty close.*

True, our final goal is to find the slope of the *tangent* line to f at $x = c$, but for now, we'll amuse ourselves by finding the slope of the *secant* line we've drawn at $x = c$. We know how to calculate the slope of a line if given two points—use the procedure from Problem 3 in Chapter 2:

$$\text{slope} = \frac{y_2 - y_1}{x_2 - x_1}$$

$$= \frac{f(c + x) - f(c)}{(c + x) - c}$$

$$= \frac{f(c + x) - f(c)}{x}$$

✏ CRITICAL POINT

Here comes the connection to limits: the smaller I make Δx, the closer the slope of the secant line comes to approximating the slope of the tangent line. I am not allowed to make $\Delta x = 0$, because that would mean I was dividing by 0 in the slope equation we created a few moments ago.

So we found the slope of the secant line, and that slope is relatively close to the slope of the tangent line we want to find—both have nearly the same incline. However, we don't want an approximation of the slope of the tangent line, we want it *exactly*. Here's the key: I am going to redraw the second point on the graph of f (the one that was Δx away from the first point), and this time I am going to make Δx smaller. Figure 8.6 shows the new, improved point and secant line. Why is it improved? It has a slope closer to the tangent line we're searching for.

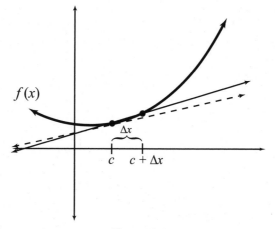

Figure 8.6

When Δx is smaller, the new point is closer to $x = c$. In addition, the solid secant line looks even more like the dotted tangent line.

This new secant line isn't as steep as the previous one, and it is an even better impersonator of the actual tangent line at $x = c$. The funny thing is, if I were to calculate its slope, it would look exactly the same as the slope I came up with before:

$$\frac{f(c + x) - f(c)}{x}$$

Here's my moment of brilliance: if I make Δx infinitely small, so small that it is basically (but not quite) 0, then the two points on the graph would be so close together that I would, in effect, actually have the tangent line. Therefore, by calculating the secant line slope, I'd actually be calculating the tangent line slope as well. How do I make Δx get that small, though? It's easy, actually, since we know limits. We're just going to find the limit of the secant slope function as Δx approaches 0. This limit is called the *difference quotient* and is the very definition of the *derivative:*

$$\lim_{x \to 0} \frac{f(c + x) - f(c)}{x}$$

CRITICAL POINT

> The formula is called the difference quotient because (1) it represents a quotient since it is a fraction, and (2) the numerator and denominator represent the difference in the y's and the x's, respectively, between the two points on our secant line.

This is the most important calculus result we have discussed thus far. We now have an admittedly ugly but very functional formula allowing us to calculate the slope of the tangent line to a function. What's amazing is that we really forced it, didn't we? We actually created a tangent line out of thin air by forcing a secant line to undergo radical and mind-altering changes. But if you're like me, you're thinking, "Enough with the theory, already—I'll probably never have to create the definition of a derivative. Instead, I'd rather know how to use the difference quotient to find a derivative." Your wish is my command.

DEFINITION

> The **derivative** of a function $f(x)$ at $x = c$ is the slope of the tangent line to f at $x = c$. You can find the value of the derivative using the **difference quotient,** which is this formula:
>
> $$\lim_{x \to 0} \frac{f(x + x) - f(x)}{x}$$
>
> As you can see, I usually write the formula with x's instead of c's, but that doesn't change the way it works.

Applying the Difference Quotient

In order to find the derivative $f'(x)$ of the function $f(x)$, you'll apply the difference quotient formula. To get the numerator, you'll plug $(x + \Delta x)$ into f and then subtract the original function $f(x)$. Then, divide that quantity by Δx, and calculate the limit of the entire fraction as Δx approaches 0.

Example 1: Use the difference quotient to find the derivative of $f(x) = x^2 - 3x + 4$, and then evaluate $f'(2)$.

Solution: The difference quotient has one ugly piece in the numerator: $f(x + \Delta x)$. So let's figure out exactly what that is ahead of time and then plug it into the formula. Remember, when you evaluate $f(x + \Delta x)$, you have to plug in $(x + \Delta x)$ into all the x terms in f. In other words, plug it into x^2 and $-3x$:

$$f(x) = x^2 - 3x + 4$$
$$f(x + \Delta x) = (x + \Delta x)^2 - 3(x + \Delta x) + 4$$
$$f(x + \Delta x) = x^2 + 2x\Delta x + (\Delta x)^2 - 3x - 3\Delta x + 4$$

That entire disgusting quantity must be substituted into the difference quotient for $f(x + \Delta x)$ now, and we'll try to simplify as much as possible:

$$\lim_{\Delta x \to 0} \frac{f(x + \Delta x) - f(x)}{\Delta x}$$

$$= \lim_{\Delta x \to 0} \frac{\left(x^2 + 2x\Delta x + (\Delta x)^2 - 3x - 3\Delta x + 4\right) - \left(x^2 - 3x + 4\right)}{\Delta x}$$

$$= \lim_{\Delta x \to 0} \frac{\cancel{x^2} + 2x\Delta x + (\Delta x)^2 \cancel{-3x} - 3\Delta x \cancel{+4} \cancel{-x^2} \cancel{+3x} \cancel{-4}}{\Delta x}$$

$$= \lim_{\Delta x \to 0} \frac{2x\Delta x + (\Delta x)^2 - 3\Delta x}{\Delta x}$$

KELLEY'S CAUTIONS

Here are the three most common errors students make when applying the difference quotient:

(1) not subtracting $f(x)$ in the numerator

(2) not distributing the negative sign through $f(x)$

(3) omitting the denominator completely

You've got to admit, that looks a lot better than it did a second ago. You were starting to panic, weren't you? All of these difference quotient problems are going to simplify significantly like this. Now, how do we evaluate the limit? Substitution is a no-go, because it results in $\frac{0}{0}$, so we should move on to the next available technique: factoring. That works like a charm:

$$\lim_{x \to 0} \frac{x(2x + x - 3)}{x}$$
$$= \lim_{x \to 0} (2x + x - 3)$$
$$= 2x + 0 - 3$$
$$= 2x - 3$$

Now for the second part of the problem: calculating $f'(2)$, the slope of the tangent line when $x = 2$. It's as easy as plugging $x = 2$ into the newfound derivative formula:

$$f'(x) = 2x - 3$$
$$f'(2) = 2(2) - 3$$
$$f'(2) = 4 - 3$$
$$f'(2) = 1$$

CRITICAL POINT

There are many notations that indicate a derivative. The most common are $f'(x)$, y', and $\frac{dy}{dx}$. The last two of these are typically used when the original function is written in "$y =$" form, rather than "$f(x) =$" form. The second derivative (the derivative of the first derivative) is denoted $f''(x)$, y'', and $\frac{d^2y}{dx^2}$.

Once you find the general derivative using the difference quotient ($f'(x) = 2x - 3$), you can then calculate any specific derivative you desire (like $f'(2)$). However, finding that general derivative is not a whole lot of fun. In fact, it's just about as fun as that time you got nothing but socks and underpants for your birthday. Calculus does offer you an alternative form of the difference quotient if you feel hatred toward this method welling up inside of you.

YOU'VE GOT PROBLEMS

Problem 1: Find the derivative of $g(x) = 5x2 + 7x - 6$ using the difference quotient and then calculate $g'(-1)$.

The Alternate Difference Quotient

I have good news and bad news for you. First, the good news: the alternate difference quotient involves much less algebra and absolutely no Δx's at all. But the bad news is that it cannot find the general derivative—you can only calculate specific values of the derivative. In other words, you'll be able to use this method to find values such as $f'(3)$, but you won't be able to find the actual derivative $f'(x)$. This definitely limits its usefulness, but it is, without question, much faster than the first method once you get used to it.

The alternate difference quotient: The derivative of $f(x)$ at the specific x-value $x = c$ can be found using the formula:

$$f'(c) = \lim_{x \to c} \frac{f(x) - f(c)}{x - c}$$

Notice the major differences between this and the previous difference quotient. For one thing, in this limit you approach the number c, not x, at which you are finding the derivative; in the other method, Δx always approached 0. In the numerator of this formula, you will calculate $f(c)$, which will be a real number; in the previous formula, both pieces of the numerator, $f(x + \Delta x)$ and $f(x)$, were functions of x. Clearly, the two formulas have different denominators as well. Since both are limits, though, evaluating them is quite similar once you've plugged in the initial values.

For grins, let's redo the second part of Example 1, since we already know the correct answer. Ever notice that math teachers just *love* doing this—reworking the same problem twice using different methods and arriving at the same answer as if by magic? I remember doing this in class, turning around at the conclusion of the second problem, and saying, "You see, they're equal!" Needless to say, I was the only one impressed. However, I still do this, hoping against hope that one day a student will faint from pure shock and delight when the answers work out the same.

Example 2: Evaluate $f'(2)$ if $f(x) = x2 - 3x + 4$. Use the difference quotient to identify $f'(x)$.

Solution: The formula requires us to know $f(c)$, in this case $f(2)$, so calculate that first:

$$f(x) = x^2 - 3x + 4$$
$$f(2) = 2^2 - 3(2) + 4$$
$$f(2) = 4 - 6 + 4$$
$$f(2) = 2$$

Now, plug that into the alternate difference quotient, and you'll be pleasantly surprised by how much simpler it looks than Example 1:

$$f'(2) = \lim_{x \to 2} \frac{f(x) - f(2)}{x-2}$$

$$= \lim_{x \to 2} \frac{(x^2 - 3x + 4) - (2)}{x-2}$$

$$= \lim_{x \to 2} \frac{x^2 - 3x + 2}{x-2}$$

To finish, evaluate the limit using the factoring method:

$$= \lim_{x \to 2} \frac{(x-2)(x-1)}{x-2}$$

$$= \lim_{x \to 2} (x - 1)$$

$$= 2 - 1$$

$$= 1$$

Like magic (although I'm sure you're unimpressed), we get the same answer as before. Ta-da!

YOU'VE GOT PROBLEMS

Problem 2: Calculate the derivative of $h(x) = \sqrt{x+1}$ when $x = 8$ using the alternate difference quotient.

The Least You Need to Know

- The slope of the tangent line to a curve at a certain point is called the derivative at that point.
- There are two forms of the difference quotient; both give the value of a function's derivative at any given x-value.
- The original form of the difference quotient can provide the general derivative formula for a function, whereas the alternate form can only give the derivative's value for a specific x.

PART 3

The Derivative

At the end of Part 2, you learned the basics of the difference quotient and that it calculates something called the derivative. In the study of calculus, the derivative is *huge.* Just about everything you do from here on out is going to use derivatives to some degree. Therefore, it's important to know exactly what they are, when they do and don't exist, and how to find derivatives of functions.

Once you've got the basic skills down, you can begin to explore the huge forest of applications that comes along with the derivative package. Since derivatives are actually rates of change, they classify and describe functions in ways you'll hardly believe. Have you ever wondered, "What's the maximum area I could enclose with a rectangular fence if one side of the rectangle is three times more than twice the other side?" If you have, well, you scare me because no one has thoughts like that. However, the good news is that you'll be able to find your answer once and for all.

CHAPTER 9

Laying Down the Law for Derivatives

One of my most memorable college professors was a kindly Korean man named Dr. Oh. One of the reasons his class sticks out in my mind is the way he was able to illustrate things with bizarre but poignant imagery. The day we first discussed the Fundamental Theorem of Calculus, he described it in his usual understated way. "Today's topic is like the day the world was created. Yesterday, not interesting. Today, interesting!"

One of the classes I took from Dr. Oh was Differential Equations. Dr. Oh constantly (but jokingly) harassed the young lady who sat next to me, because she would always do things the long way. No matter what shortcuts we learned, she wouldn't use them. I never understood why, and she simply explained to me, "This is the way I do things. I can't change it now!" I remember Dr. Oh repeatedly asking her, "If you want potatoes, do you buy a farm, till the field, plant the seed, nurture the plants, and then harvest the potatoes? If I were you, I would just go to the grocery store."

In the land of derivatives, the difference quotient (see Chapter 8) is the equivalent of growing your own potatoes. Sure, the process works, but I gave you very specific examples so that it would work for you without any trouble or heartache. I was shielding you against the harsh weather of

In This Chapter

- When can you find a derivative?
- Calculating rates of change
- Simple derivative techniques
- Derivatives of trigonometric functions
- Multiple derivatives
- Using calculators to evaluate derivatives

complicated derivatives to come. However, I have to let you grow up sometime and stare in the face of an ugly, complicated derivative. The good news, though, is that you can buy all your solutions from the grocery store.

When Does a Derivative Exist?

Before you run around finding derivatives willy-nilly, you should know that there are three specific instances in which the derivative to a function fails to exist. Even if you get a numerical answer when calculating a derivative, it's possible that the answer is invalid, because there actually is no derivative! Be extra cautious if the graph of your function contains any of the following things.

> **CRITICAL POINT**
>
> You'll hear the statement "Differentiability implies continuity" in your calculus class. That means exactly this: if a function has a derivative at a specific x value, then the function *must* also be continuous at that x value. That statement is the logical equivalent of saying, "If a function is *not* continuous at a certain point, then that function is not differentiable there either."

Discontinuity

A derivative cannot exist at a point of discontinuity. It doesn't matter if the discontinuity is removable or not. If a function is discontinuous at a specific x value, there cannot be a derivative there. For example, take a look at this function:

$$f(x) = \frac{(x-1)(x+2)}{(x+2)(x-6)}$$

Without doing a bit of work, you can conclude that $f(x)$ has no derivative at $x = -2$ and $x = 6$. In other words, $f(x)$ is not *differentiable* at those values of x.

Sharp Point in the Graph

If a graph contains a sharp point (also known as a cusp), then the function has no derivative at that point. Not many functions have cusps; in fact, they are pretty rare. You're most likely to see them in functions containing absolute values and in piecewise-defined functions whose pieces meet, but not smoothly. In Figure 9.1, you'll find the graphs of the function $f(x) = |x-1| - 2$ and this piecewise-defined function:

$$g(x) = \begin{cases} x^2, & x \leq 1 \\ x, & x > 1 \end{cases}$$

Both $f(x)$ and $g(x)$ contain *nondifferentiable* cusps at $x = 1$.

DEFINITION

A function is **differentiable** at a given value of x if you can take the derivative of the function at that x value. In other words, $f(x)$ is differentiable at $x = c$ if $f'(c)$ exists. A function whose derivative does not exist at a specific x value is said to be **nondifferentiable** there.

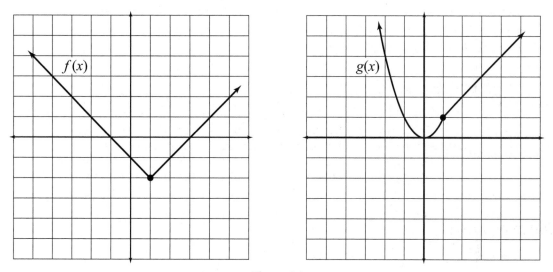

Figure 9.1
Both graphs have a sharp point at x $+$ *1 and, therefore, are not differentiable there.*

Vertical Tangent Line

Remember that the derivative is defined as the slope of the tangent line. What if the tangent line is vertical? Keep in mind that vertical lines don't have a slope, so a derivative cannot exist there. It's pretty tough to spot when this happens using only a graph, but luckily, the mathematics of derivatives is quick to expose it when it happens, as shown in the next example.

Example 1: Show that no derivative exists for the function $f(x) = x^{1/3}$ when $x = 0$.

Solution: You don't know how to find the derivative of $f(x) = x^{1/3}$ yet (but you will soon), so I'll tell you that it's $f'(x) = \frac{1}{3x^{2/3}}$. If you try to evaluate $f'(0)$, you get:

$$f'(0) = \frac{1}{3(0)^{2/3}}$$

$$= \frac{1}{3 \cdot 0}$$

$$= \frac{1}{0}$$

The slope of the tangent line is a nonexistent number, because you can't divide by 0.

Basic Derivative Techniques

Learning how to find derivatives using the difference quotient can be long, tedious work, but once you've mastered it, you've "paid your dues," so to speak. Now, you'll learn some really handy techniques. Here are three derivative shortcuts that will make things a whole lot faster and easier.

The Power Rule

Even though the Power Rule can only find very basic derivatives, you'll definitely use it more than any other of the rules we'll learn. In fact, it often pops up in the final steps of other rules, but let's not get ahead of ourselves. Any term in the form ax^n can be differentiated using the Power Rule.

> ✏ **CRITICAL POINT**
>
> Don't worry about the phrase "with respect to x" in the Power Rule definition. Because x is the only variable in the expression, we really don't need to say that. In Chapter 10, we'll cover implicit differentiation, and then you'll have to know what the phrase "with respect to" actually means. For now, just understand that the phrase refers to the variable in the problem, but won't affect any of your derivative techniques.

The Power Rule: The derivative of the term ax^n (with respect to x), where a and n are real numbers, is $(a \cdot n) x^{n-1}$.

Here are the steps you'll use to find a derivative with the Power Rule:

1. Multiply the coefficient by the variable's exponent. If no coefficient is stated—in other words, the coefficient equals 1—the exponent becomes the new coefficient.

2. Subtract 1 from the exponent.

Some examples will shed some light on the matter.

Example 2: Use the Power Rule to find the derivative of $f(x) = \frac{4}{3}x^3 + 6x - 5$.

Solution: Even though there are a number of terms here, you can find the derivative of each one separately using the Power Rule. Before you start, I'll tell you that the derivative of the constant term (–5) is 0. (More on this in a minute.) For the other terms, multiply the coefficient of each one by the exponent and then subtract 1 from the exponent:

$$f'(x) = \left(\frac{4}{3} \cdot 3\right)x^{3-1} + \left(6 \cdot 1\right)x^{1-1} - 0$$
$$= 4x^2 + 6x^0$$
$$= 4x^2 + 6$$

Remember, the exponent of $6x$ is understood to be 1 because it's not written explicitly, and a variable to the zero power equals 1: $6x^0 = 6 \cdot 1 = 6$.

The derivative of any constant is 0. Here is a quick justification if you are interested. Consider the constant function $g(x) = 7$. If you wanted to, you could write this function with a variable term: $g(x) = 7x^0$. I am not changing the value of the function because a variable to the 0 power has a value of 1, and $7 \cdot 1 = 7$. So now that you've rewritten the function, use the Power Rule:

$$g(x) = 7x^0$$
$$g'(x) = \left(7 \cdot 0\right)x^{0-1}$$
$$g'(x) = 0x^{-1}$$
$$g'(x) = 0$$

YOU'VE GOT PROBLEMS

Problem 1: Find derivatives using the Power Rule:

(a) $y = \frac{2}{3}x^3 + 3x^2 - 6x + 1$

(b) $f(x) = \sqrt[3]{x} + 2\sqrt[5]{x}$

The Product Rule

If a function contains two variable expressions multiplied together, you cannot simply find the derivative of each and multiply the results. For example, the derivative of $x^2 \times (x^3 - 3)$ is *not* $(2x)(3x^2)$. Instead, you have to use a very simple formula, which (by the way) you should memorize.

> **KELLEY'S CAUTIONS**
>
> Overlooking the Product Rule is a very common mistake in calculus. Remember: if two variable expressions are multiplied together, you *have* to use the Product Rule. If, however, you want to find the derivative of $5 \times 7x^2$, you don't need the Product Rule (because 5 is not a variable expression). Instead, you can rewrite it as $35x^2$ and use the Power Rule to get the correct derivative of $70x$.

The Product Rule: If a function $h(x) = f(x) \times g(x)$ is the product of two differentiable functions $f(x)$ and $g(x)$, then

$$h'(x) = f(x) \cdot g'(x) + f'(x) \cdot g(x)$$

Here's what that means. If a function is created by multiplying two other functions together, then the derivative of the overall function is the first one times the derivative of the second plus the second one times the derivative of the first.

Example 3: Differentiate $f(x) = (x^2 + 6)(2x - 5)$ using (1) the Product Rule, and (2) the Power Rule, and show that the results are equal. *Hint:* to use the Power Rule, you'll first have to multiply the terms together.

Solution: (1) Apply the Product Rule:

$$f'(x) = \left(x^2 + 6\right)(2) + (2x)(2x - 5)$$
$$= 2x^2 + 12 + 4x^2 - 10x$$
$$= 6x^2 - 10x + 12$$

(2) As the hint indicates, you need to multiply those binomials together before you can apply the Power Rule: $f(x) = 2x^3 - 5x^2 + 12x - 30$. Now, apply the Power Rule to get $f'(x) = 6x^2 - 10x + 12$, which matches the answer from part (1).

> **YOU'VE GOT PROBLEMS**
>
> Problem 2: Find the derivative of $g(x) = (2x - 1)(x + 4)$ using the Power Rule and the Product Rule, and show that the results are the same.

The Quotient Rule

Just as the Product Rule prevents you from simply taking individual derivatives when you're multiplying, the Quotient Rule prevents the same for division. Every year on my first derivatives exam, one of the problems is to find the derivative of something like $\frac{x^2+7x}{3x^3+2x+4}$, and half of my students always answer $\frac{2x+7}{9x^2+2}$, no matter how many times I warn them to use the Quotient Rule. You *must* use the Quotient Rule any time two variable expressions are divided.

KELLEY'S CAUTIONS

It is very important to get the subtraction order correct in the numerator of the Quotient Rule. Whereas in the Product Rule, either of the two functions could be f or g, in the Quotient Rule, g must be the denominator of the function.

The Quotient Rule: If $h(x)=\frac{f(x)}{g(x)}$, where $f(x)$ and $g(x)$ are differentiable functions and $g(x) \neq 0$, then:

$$h'(x)=\frac{g(x)\cdot f'(x)-f(x)\cdot g'(x)}{\left[g(x)\right]^2}$$

In other words, to find the derivative of a fraction, take the bottom times the derivative of the top and subtract the top times the derivative of the bottom; divide all of that by the bottom squared. Of course, by top and bottom, I mean numerator and denominator, respectively.

Example 4: Find the derivative of $y=\frac{3x+7}{x^2-1}$ using the Quotient Rule.

Solution: The numerator is $f(x)$ in the Quotient Rule, and the denominator is $g(x)$: $f(x) = 3x + 7$ and $g(x) = x^2 - 1$. Therefore, $f'(x) = 3$ and $g'(x) = 2x$. Plug all of these values into the appropriate spots in the Quotient Rule:

$$y'=\frac{g(x)\cdot f'(x)-f(x)\cdot g'(x)}{\left[g(x)\right]^2}$$
$$=\frac{\left(x^2-1\right)\cdot 3-(3x+7)\cdot 2x}{\left(x^2-1\right)^2}$$
$$=\frac{3x^2-3-6x^2-14x}{x^4-2x^2+1}$$
$$=\frac{-3x^2-14x-3}{x^4-2x^2+1}$$

YOU'VE GOT PROBLEMS

Problem 3: Use the Quotient Rule to differentiate $f(x)=\frac{3x^4+2x^2-7x}{x-5}$ and simplify $f'(x)$.

The Chain Rule

Consider, for a moment, the functions $f(x) = \sqrt{x}$ and $g(x) = 3x + 1$. With the skills you now possess, you could find the derivative of each using the Power Rule. You could even find the derivatives of their product $f(x) \cdot g(x)$ or their quotient $\frac{f(x)}{g(x)}$, using the Product and Quotient Rules, respectively (no big surprise there).

However, you don't know how to find the derivative of two functions plugged into (or "composed with") one another. In other words, the derivative of $f(g(x)) = \sqrt{3x+1}$ requires a technique you've not yet learned, a technique called the Chain Rule.

If this function were simpler, such as $y = \sqrt{x}$, there would be no need for the Chain Rule, but the inner function (in this case $3x + 1$) is too complicated. Here's a good rule of thumb: if a function contains something other than a single variable, like x, then you should use the Chain Rule to find its derivative.

The Chain Rule: Given the composite function $h(x) = f(g(x))$, where $f(x)$ and $g(x)$ are differentiable functions, then $h'(x) = f'(g(x)) \cdot g'(x)$.

CRITICAL POINT

The derivatives of logarithmic and exponential equations use the Chain Rule heavily. Make sure to learn these patterns:

- $\frac{d}{dx}\left(\log_a f(x)\right) = \frac{1}{(\ln a) \cdot f(x)} \cdot f'(x)$

- $\frac{d}{dx}\left(a^{f(x)}\right) = (\ln a) \cdot a^{f(x)} \cdot f'(x)$

There are special cases for the natural logarithm ($\ln x$) and the natural exponential function (e^x), so you'll see those more often: $\frac{d}{dx}\left(\ln x\right) = \frac{1}{x}$ and $\frac{d}{dx}\left(e^x\right) = e^x$.

In other words, to take the derivative of an expression where one function is "trapped inside" another function, you follow these steps:

1. Take the derivative of the "outer" function, leaving the trapped, "inner" function alone.

2. Multiply the result by the derivative of the "inner" function.

Example 5: Use the Chain Rule to find the derivative of $y = \sqrt{3x+1}$.

Solution: Rewrite the function so that it's clear what's actually plugged into what. In this case, $3x + 1$ is plugged into $\sqrt{x}$. In other words, if $f(x) = \sqrt{x}$ (the outer function) and $g(x) = 3x + 1$ (the inner function, because it's trapped inside the square root symbol in $f(x)$), then $f\left(g\left(x\right)\right)=\sqrt{3x+1}$. Rewriting the function like this helps you plug everything into the right spots in the Chain Rule formula.

Your first step is to take the derivative of $f(x)$, leaving $g(x)$ alone. This just means you should find the derivative of $f(x)$ and, once you're done, plug $g(x)$ into all of its x spots. According to the Power Rule, if $f(x) = x^{1/2}$:

$$f'\left(x\right)=\frac{1}{2}x^{-1/2}$$

$$=\frac{1}{2x^{1/2}}$$

$$=\frac{1}{2\sqrt{x}}$$

Now plug $g(x)$ in for x:

$$f'\left(g\left(x\right)\right)=\frac{1}{2\sqrt{3x+1}}$$

You're almost done. The final step is to multiply this ugly monstrosity of a fraction by the derivative of $g(x)$. A quick nod to the Power Rule tells you that if $g(x) = 3x + 1$, then $g'(x) = 3$:

$$y'=f'\left(g\left(x\right)\right)\cdot g'\left(x\right)$$

$$=\frac{1}{2\sqrt{3x+1}}\cdot 3$$

$$=\frac{3}{2\sqrt{3x+1}}$$

YOU'VE GOT PROBLEMS

Problem 4: Use the Chain Rule to differentiate $y = (x^2 + 1)^5$.

Rates of Change

Derivatives are so much more than what they seem. True, they give the slope of the tangent line to a curve. But that slope can tell us a great deal about the curve. One characteristic of the derivative we will exploit time and time again is this: *the derivative of a curve tells us the instantaneous rate of change of the curve*. This is key because a "curvy" function will change at different rates throughout its domain.

On the other hand, the graph of a line always changes at the exact same rate. For example, $g(x) = 4x - 3$ will always increase at a rate of 4, because that is the slope of the line and also the derivative. Curves, however, do not have the same slope everywhere, so we rely on the slopes of their tangent lines. Sometimes a curve is increasing quickly and the tangent line is steep (causing a high-valued derivative). At other places the curve may be increasing shallowly or even decreasing, causing the derivative to be small or negative, respectively. Look at the graph of $f(x)$ in Figure 9.2.

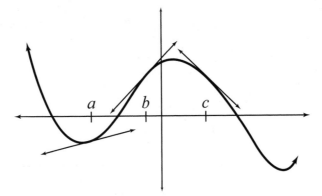

Figure 9.2
The graph of f(x)*, with three points of interest shown.*

At $x = a$, f is increasing ever so slightly, causing the tangent line there to be shallow. Because a shallow line has a slope close to 0, the derivative here will be very small. In other words, the rate of change of the graph is very small at the instant that $x = a$. However, at the instant that $x = b$, the graph is climbing more rapidly, causing a steeper tangent line, which in turn causes a larger derivative. Finally, at $x = c$, the graph is decreasing, so the instantaneous rate of change there is negative (because the slope of the tangent line is negative).

You can also use the slope of a *secant* line to determine rates of change on a graph. However, the slope of a secant line describes something different: the *average* rate of change over some portion of the graph. Finding the slope of a secant line is very easy, as you'll see in the next example.

CRITICAL POINT

Remember, the slope of a tangent line to a curve tells you the curve's rate of change at that value of x (the instantaneous rate of change, because you can only tell what's going on at that instant). The slope of the secant line to a curve tells you the *average* rate of change over the specified interval.

Example 6: Poteet, Inc. has just introduced a new, revolutionary brand of athletic sock into the market. The new innovation is a special sweat-absorbing "cotton-esque" material that supposedly prevents foot odor. On their fourth day of sales, the snappy slogan, "If you smell feet, they ain't wrapped in Poteet's," was released, and sales immediately increased. Figure 9.3 is a graph of the number of units sold during the first six days of sales.

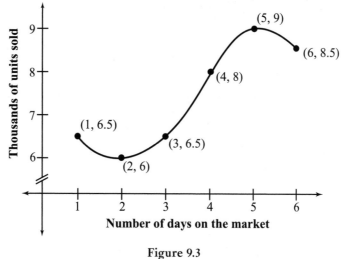

Figure 9.3
The rise of a new sweat sock empire.

What was the average rate of units sold per day between day one and day six?

Solution: The problem asks us to find an average rate of change, which translates to finding the slope, *m,* of the secant line connecting the points (1,6.5) and (6,8.5). To do that, use our tried-and-true method from algebra:

$$m = \frac{y_2 - y_1}{x_2 - x_1}$$
$$= \frac{8.5 - 6.5}{6 - 1}$$
$$= \frac{2}{5}$$

That means Poteet's socks sold at a rate of two-fifths of a thousand units per day, or $\frac{2}{5}(1,000) = 400$ units/day on average. So even through the moderate decreases they experienced, the new slogan probably helped.

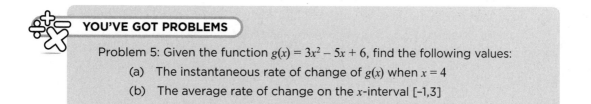

YOU'VE GOT PROBLEMS

Problem 5: Given the function $g(x) = 3x^2 - 5x + 6$, find the following values:

(a)   The instantaneous rate of change of $g(x)$ when $x = 4$

(b)   The average rate of change on the x-interval [–1,3]

Trigonometric Derivatives

Before we leave the land of simple derivatives, we must first discuss trigonometric derivatives. Each trig function has a unique derivative that you should memorize. Whereas some are easy to build from scratch (as you'll see in Problem 6), others are quite difficult, so it's best to memorize the entire list. Trust me—a little memorization now goes a long way later. Take a deep breath and gaze upon the following list of important trig derivatives:

$$\frac{d}{dx}\left(\sin x\right) = \cos x \qquad\qquad \frac{d}{dx}\left(\arcsin x\right) = \frac{1}{\sqrt{1-x^2}}$$

$$\frac{d}{dx}\left(\cos x\right) = -\sin x \qquad\qquad \frac{d}{dx}\left(\arccos x\right) = \frac{-1}{\sqrt{1-x^2}}$$

$$\frac{d}{dx}\left(\tan x\right) = \sec^2 x \qquad\qquad \frac{d}{dx}\left(\arctan x\right) = \frac{1}{1+x^2}$$

$$\frac{d}{dx}\left(\cot x\right) = -\csc^2 x \qquad\qquad \frac{d}{dx}\left(\operatorname{arccot} x\right) = \frac{-1}{1+x^2}$$

$$\frac{d}{dx}\left(\sec x\right) = \sec x \tan x \qquad\qquad \frac{d}{dx}\left(\operatorname{arcsec} x\right) = \frac{1}{|x|\sqrt{x^2-1}}$$

$$\frac{d}{dx}\left(\csc x\right) = -\csc x \cot x \qquad\qquad \frac{d}{dx}\left(\operatorname{arccsc} x\right) = \frac{-1}{|x|\sqrt{x^2-1}}$$

It's not as bad as you think—half of the inverse trig derivatives are different from the other half by only a negative sign.

You'll notice that I have included the inverse trig functions in this list, but you may not recognize them. Instead of using the notation $y = \sin^{-1}x$ to indicate the inverse sine, I use the notation $y = \arcsin x$. I am a huge fan of the latter notation, because $\sin^{-1}x$ looks a lot like $(\sin x)^{-1}$, which is equal to $\csc x$ (which is not the inverse of $\sin x$).

> ✏️ **CRITICAL POINT**
>
> The notation $\frac{d}{dx}\left(f(x)\right)$ means "take the derivative of the expression inside the parentheses." In other words, $\frac{d}{dx}\left(f(x)\right) = f'(x)$.

You'll have to be able to use these formulas with the Product, Quotient, and Chain Rules, so here are a couple of examples to get you used to them. Remember, if a trig function contains anything except a single variable (like x), you have to use the Chain Rule to find the derivative.

Example 7: If $f(x) = \cos x \sin 2x$, identify $f'(x)$ and evaluate $f'\left(\frac{\pi}{2}\right)$.

Solution: Because this function is the product of two variable expressions, you'll have to use the Product Rule. In addition, you'll have to use the Chain Rule to differentiate $\sin 2x$, because it contains more than just x inside the sine function. According to the Chain Rule, $\frac{d}{dx}\left(\sin 2x\right) = \cos\left(2x\right) \cdot 2 = 2\cos 2x$. Here's the Product Rule in action:

$$f'(x) = \cos x \cdot 2\cos 2x + (-\sin x)(\sin 2x)$$
$$f'(x) = 2\cos x \cos 2x - \sin x \sin 2x$$
$$f'\left(\tfrac{\pi}{2}\right) = 2\cos\left(\tfrac{\pi}{2}\right)\cos\left(\tfrac{2\pi}{2}\right) - \sin\left(\tfrac{\pi}{2}\right)\sin\left(\tfrac{2\pi}{2}\right)$$
$$= 2(0)(-1) - 1(0)$$
$$= 0 - 0$$
$$= 0$$

YOU'VE GOT PROBLEMS

Problem 6: Use the Quotient Rule to prove that $\frac{d}{dx}(\cot x) = -\csc^2 x$.

Tabular and Graphical Derivatives

This chapter is chock full of important techniques and formulas to memorize and practice, but as you're mastering the Power, Product, Quotient, and Chain Rules, don't get so caught up in the calculations that you miss the concepts. Derivatives are a foundational element of calculus, so let's throw a few nontraditional examples at you before we conclude the chapter.

Example 8: Assume functions $f(x)$ and $g(x)$ are continuous and differentiable for all real numbers. The following table lists values of the functions and their derivatives for specific x values.

x	$f(x)$	$g(x)$	$f'(x)$	$g'(x)$
−3	0	−8	2	−3
−2	3	6	−5	−6
−1	1	2	−6	−10
0	−3	−2	7	12
1	−2	−1	−8	−4
2	−1	7	1	5
3	−4	3	−9	−15

Based on this information, calculate the following:

(a) $h'(1)$, given $h(x) = f(x) \cdot g(x)$ (b) $j'(-3)$, given $j(x) = g(f(x))$

Solution: Although you don't have specific functions that define $f(x)$ and $g(x)$, you can still apply the derivative techniques you learned in this chapter.

(a) The function $h(x)$ is defined as the product of functions $f(x)$ and $g(x)$. That means you need to use the Product Rule to calculate $h'(x)$:

$$h(x) = f(x) \cdot g(x)$$
$$h'(x) = f(x) \cdot g'(x) + g(x) \cdot f'(x)$$

What now? You don't know what $f(x)$ or $g(x)$ are equal to, so how are you supposed to know what their derivatives are? No need to worry. You aren't asked to find the general derivative, just the value of the derivative when $x = 1$.

$$h'(1) = f(1) \cdot g'(1) + g(1) \cdot f'(1)$$

Find $f(1), f'(1), g(1),$ and $g'(1)$ in the table and substitute them into the formula.

$$h'(1) = (-2)(-4) + (-1)(-8)$$
$$= 8 + 8$$
$$= 16$$

(b) Function $j(x)$ is a composition of functions: $f(x)$ is substituted into $g(x)$. You need to apply the Chain Rule to calculate $j'(x)$:

$$j(x) = g(f(x))$$
$$j'(x) = g'(f(x)) \cdot f'(x)$$

You're asked to calculate $j'(-3)$, so replace x with -3:

$$j'(-3) = g'(f(-3)) \cdot f'(-3)$$

According to the chart, $f(-3) = 0$.

$$j'(-3) = g'(0) \cdot f'(-3)$$

Once again, refer to the chart to identify the values of $g'(0)$ and $f'(-3)$:

$$j'(-3) = 12 \cdot 2$$
$$= 24$$

 YOU'VE GOT PROBLEMS

Problem 7: Use the table of values provided in Example 8 to calculate $k'(-1)$, given $k(x) = \dfrac{f(x)}{g(x)}$.

Example 9: Given the graph of $f(x)$ in Figure 9.4, estimate the following:

(a) $f'(3)$ (b) $f'(-2)$

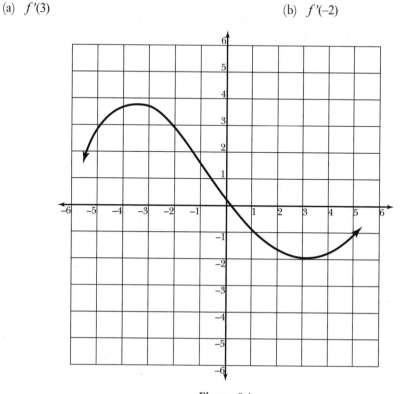

Figure 9.4
The graph of a function f(x).

Solution: As Chapter 8 explains, the derivative of a function is the slope of the tangent line at a specific x value. Therefore, to estimate the derivatives of $f(x)$ in this example, you should draw tangent lines at the given x values and calculate their slopes.

(a) The graph of $f(x)$ seems to be changing direction at $x = 3$. Before it passes $x = 3$, as you travel from left to right, it is decreasing. However, after $x = 3$, it increases. The tangent line at $x = 3$, therefore, will be horizontal, as illustrated in Figure 9.5. Remember, this is an estimation, so even if the tangent isn't perfectly horizontal, this is a good enough guess.

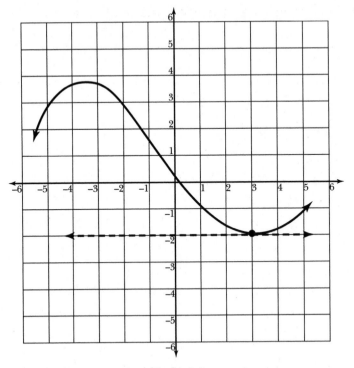

Figure 9.5
The line tangent to f(x) *when* x = 3 *is horizontal.*

What is the derivative when $x = 3$? The slope of a horizontal line is 0, so you can confidently assert that $f'(3) \approx 0$.

(b)   Draw the line tangent to the graph at $x = -2$.

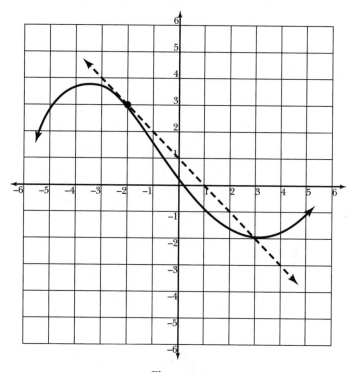

Figure 9.6
The line tangent to the graph of f(x) *when* x $= -2$.

Your line may vary a little from mine, but remember that you're only asked for an estimate. My dotted tangent line passes through points $(-2,3)$ and $(1,0)$. Use the slope formula to calculate the slope of the tangent:

$$\text{slope} = \frac{y_2 - y_1}{x_2 - x_1}$$

$$= \frac{0-3}{1-(-2)}$$

$$= \frac{-3}{1+2}$$

$$= -\frac{3}{3}$$

$$= -1$$

You conclude that $f'(-2) \approx -1$.

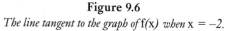

YOU'VE GOT PROBLEMS

Problem 8: Use the graph of $f(x)$ in Figure 9.4 to estimate $f'(4)$.

Technology Focus: Calculating Derivatives

Different calculators have different capabilities when it comes to calculating derivatives. Some calculators, like the TI-89, can differentiate symbolic expressions. By accessing the Calculus menu on the home screen, you can select the differentiate function, as illustrated in Figure 9.7. If you own a symbolic calculator, you can check most of your homework in a snap. (However, don't be tempted to depend on your calculator to do your work for you.)

Figure 9.7

The Calc menu contains common calculus functions, including "differentiate."

For example, you can check your work for Example 3 in this chapter by entering the expression, as shown in Figure 9.8. Notice that the expression is followed by ",x" indicating that the expression you typed contains the variable *x*, and that is the variable you are differentiating with respect to. (More about what "with respect to" means in Chapter 10.)

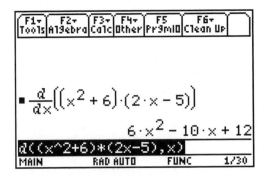

Figure 9.8

The calculator verifies the solution to Example 3 in this chapter, from the Product Rule section.

Pretty powerful! The calculator applied the Product Rule for you automatically and even simplified the expression. Symbolic calculators are very smart devices. Check out Figure 9.9 to see that even trigonometric and inverse trigonometric functions are just as easy to differentiate.

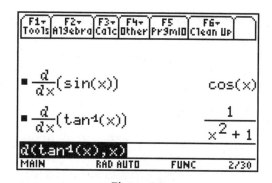

Figure 9.9
The TI-89 doesn't blink an electronic eye at the thought of differentiating trigonometric functions.

Here's the bad news: symbolic calculators are so powerful that most classes and exams don't allow them. After all, teachers and exam makers want to make sure *you* understand calculus; they aren't interested in measuring your graphing calculator skills. Other calculators are allowed, including the TI-84 family. While they cannot handle symbolic differentiation, they can calculate derivatives at specific *x* values.

For example, in Example 3 (and in the calculator screens above), you determined that the derivative of $f(x) = (x^2 + 6)(2x - 5)$ was $6x^2 - 10x + 12$. Your TI-84 calculator would not be able to tell you the derivative expression, but it could *evaluate* the derivative for a specific value of *x*. In other words, it can't tell you what $f'(x)$ is, but it can tell you what $f'(-2)$ is ... almost.

Press the MATH button and scroll down to the "nDeriv(" option, as illustrated in Figure 9.10. This stands for "numeric derivative," as opposed to symbolic derivative.

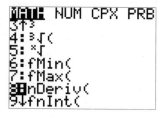

Figure 9.10
Some of your screens may look different from mine if you have MathPrint enabled in the Mode menu.
You can always toggle display options in the Mode menu.

Now type the expression from Example 3, followed by ",x,–2)" to indicate that you are taking the derivative of the x variable and evaluating the derivative at x = –2. See Figure 9.11 for the result.

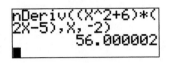

Figure 9.11

This is an estimate of the correct answer. It is not 100 percent accurate.

We can give the calculator a lot of credit for trying, but let's be honest—it did not get the answer right. If you substitute x = –2 into the derivative you get 56, not 56.000002.

$$f'(x) = 6x^2 - 10x + 12$$
$$f'(-2) = 6(-2)^2 - 10(-2) + 12$$
$$f'(-2) = 6(4) - 10(-2) + 12$$
$$f'(-2) = 24 + 20 + 12$$
$$f'(-2) = 56$$

Why does the calculator get it wrong? Remember, it's not actually calculating the symbolic derivative. It is using a method similar to the difference quotient to calculate a tangent slope, so you cannot count on it for exact measurements. However, 56.000002 is pretty close to 56, so you can still use it to check any numeric derivative answers.

One warning: your calculator may try to calculate a derivative *even if it doesn't exist!* For example, the function $g(x) = |x|$ is not differentiable when x = 0, because it contains a cusp there. The algorithm your calculator uses to approximate the derivative may not test for cusps, so the calculator reports a derivative of 0 (see Figure 9.12).

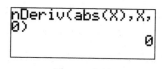

Figure 9.12

The derivative of $y = |x|$ does not exist when x = 0, but that's news to your calculator, which managed to find a derivative anyway.

There are pros and cons to using your graphing calculator. A symbolic calculator is very smart—too smart, you could argue, to use on a quiz or test. A calculator that provides only numeric derivatives is smart enough to be helpful, as long as you interpret the results appropriately.

The Least You Need to Know

- If a function is differentiable, it must also be continuous.
- A function is not differentiable at a point of discontinuity, a sharp point (cusp), or where the tangent line is vertical.
- The slope of a function's tangent line gives its instantaneous rate of change, and the slope of its secant line gives the average rate of change.
- Products and quotients of variable expressions must be differentiated using the Product and Quotient Rules, respectively.
- You must use the Chain Rule to differentiate any function that contains something other than just x.
- You can use your calculator to check your work when evaluating derivatives.

CHAPTER 10

Common Differentiation Tasks

Even though the derivative is just the slope of a tangent line, its uses are innumerable. We've already seen that it describes the instantaneous rate of change of a nonlinear function. However, that hardly explains why it's one of the most revolutionary mathematical concepts in history. Soon we'll be exploring more (and substantially more exciting) uses for the derivative.

In the meantime, there's a little bit more grunt work to be done. (That makes you happy to read, doesn't it?) This chapter will help you perform specific tasks and find derivatives for very particular situations. Think of learning derivatives as being like trying to get your body in shape. In the last chapter, you learned the basics, the equivalent of a good cardiovascular workout, working all of your muscles in harmony with each other. In this chapter, we're working out specific muscle groups, one section at a time. There's not a lot of similarity between each individual topic here, but exercising all of these abilities at the appropriate time (and knowing when that time arrives) is essential to getting yourself in shape mathematically.

In This Chapter

- Equations of tangent and normal lines
- Differentiating equations containing multiple variables
- Derivatives of inverse functions
- Differentiating parametric equations
- Solving gross equations with your calculator

Finding Equations of Tangent Lines

Writing tangent line equations is one of the most basic and foundational skills in calculus. You already know how to create the equation of a line using point-slope form (see Chapter 2). Since it's the equation of a tangent line you're after, the slope is the derivative of the function! All that's left to do is figure out the appropriate point, and if that were any easier, it'd be illegal.

Example 1: Write the equation of the tangent line to the curve $f(x) = 3x^2 - 4x + 1$ when $x = 2$.

Solution: Take a look at the graph of $f(x)$ in Figure 10.1 to get a sense of our task.

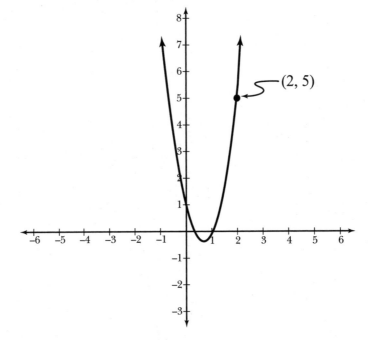

Figure 10.1

The graph of $f(x) = 3x^2 - 4x + 1$ *and a future point of tangency.*

You want to find the equation of the tangent line to the graph at the indicated point (when $x = 2$). This is the point of tangency, where the tangent line will strike the graph. Therefore, this point is both on the curve and on the tangent line. Since point-slope form requires you to know a point on the line in order to create the equation of that line, you'll need to know the coordinates of this point. Since you already know the x value, plug it into $f(x)$ to get the corresponding y value:

$$f(2) = 3(2)^2 - 4 \cdot 2 + 1$$
$$= 12 - 8 + 1$$
$$= 5$$

So the point (2,5) is on the tangent line. Now all you need is the slope of the tangent line, f' (2):

$$f'(x) = 6x - 4$$
$$f'(2) = 6(2) - 4$$
$$f'(2) = 12 - 4$$
$$f'(2) = 8$$

Now that you know a point on the tangent line and the correct slope, slap those values into point-slope form and out pops the correct tangent line equation:

$$y - 5 = 8(x - 2)$$
$$y - 5 = 8x - 16$$
$$y = 8x - 11$$

Figure 10.2 verifies the solution visually.

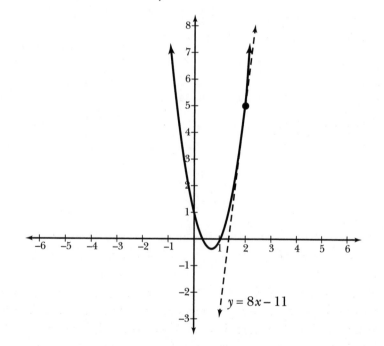

Figure 10.2
The (dotted) line y = 8x − 11 *is tangent to* f(x) *at* x = 2.

YOU'VE GOT PROBLEMS

Problem 1: Find the equation of the tangent line to $g(x) = 3x^3 - x^2 + 4x - 2$ when $x = -1$.

Occasionally you'll be asked to find the equation of the *normal line* to a curve. Because the normal line is perpendicular to the tangent line at the point of tangency, you use the same point to create the normal line, but the slope of the normal line is the negative reciprocal of the slope of the tangent line.

DEFINITION

A **normal line** is perpendicular to a function's tangent line at the point of tangency.

Example 2: Calculate the slope of the normal line to the curve $g(x) = \tan(x^2 - \pi)$ at $x = \frac{\pi}{4}$, reporting the answer accurate to the thousandths place.

Solution: Now *this* is a crazy-looking graph. I definitely do not recommend trying to sketch it by hand. If you don't mind spoilers, go ahead and peek at Figure 10.3. Anyway, enough gawking. Let's keep moving. Remember, the slope of the normal line is perpendicular to the slope of the tangent line. Apply the Chain Rule to calculate the derivative of $g(x)$:

$$g'(x) = \sec^2\left(x^2 - \pi\right) \cdot \frac{d}{dx}\left(x^2 - \pi\right)$$
$$= \sec^2\left(x^2 - \pi\right) \cdot 2x$$
$$= 2x \sec^2\left(x^2 - \pi\right)$$

Now calculate $g'\left(\frac{\pi}{4}\right)$. It is going to get messy, so plan on using decimal forms provided by your calculator.

$$g'\left(\frac{\pi}{4}\right) = 2\left(\frac{\pi}{4}\right)\sec^2\left[\left(\frac{\pi}{4}\right)^2 - \pi\right]$$
$$= \frac{\pi}{2} \cdot \sec^2\left(\frac{\pi^2}{16} - \pi\right)$$
$$\approx 2.3607725$$

Although that number was a little ugly looking, you're headed in the right direction. This is the perfect time to double-check your derivative with your calculator, as demonstrated in Figure 10.3.

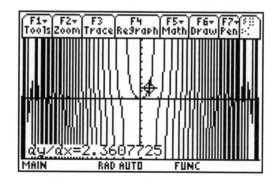

Figure 10.3
If decimals bring you no joy, I'd argue that knowing the actual value of the derivative,
$$\frac{\pi}{2} \cdot \sec^2\left(\frac{\pi^2}{16} - \pi\right), \textit{isn't much better.}$$

The slope, m, of the normal line to $g(x)$ at $x = \frac{\pi}{4}$ is the opposite reciprocal of the slope of the tangent line:

$$m \approx -\left(\tfrac{1}{2.3607725}\right)^{-1}$$
$$\approx -0.424$$

Implicit Differentiation

I've mentioned the phrase "with respect to x" a few times in other chapters, but now I need to define exactly what that means. In 95 percent of your problems in calculus, the variables in your expression will match the variable you are "respecting" in that problem. For example, the derivative of $5x^3 + \sin x$, with respect to x, is $15x^2 + \cos x$. The fact that I said you were finding the derivative with respect to x didn't make the problem any harder or any different. In fact, I didn't have to tell you which variable you were "respecting," so to speak, because x was the only variable in the problem.

In this section, we'll take the derivative of equations containing x and y, and I will always ask you to find the derivative with respect to x. What is the derivative of y with respect to x, you ask? The answer is this notation: $\frac{dy}{dx}$. It is literally read, "the derivative of y with respect to x." The numerator tells you what you're deriving, and the denominator tells you what you're respecting.

Let's try a slightly more complex derivative. What is the derivative of $3y^2$, with respect to x? The first thing to notice is that the variable in the expression does not match the variable you're respecting, so you treat the y as a completely separate function and apply the Chain Rule. I know you're not used to using the Chain Rule when there's only a single variable inside the function, but if that variable is not the variable you're respecting, you have to give it a hard time and "rough it up" a little. So to differentiate $3y^2$, start by deriving the outer function and leaving y

(the inner function) alone to get $6y$. Now multiply this by the derivative of y with respect to x, and you get:

$$6y \cdot \frac{dy}{dx}$$

You will encounter odd derivatives like this whenever you cannot solve an equation for y or for $f(x)$. You may not have noticed, but every single derivative question until now has been worded "Find the derivative of $y = \ldots$" or "Find the derivative of $f(x) \ldots$." When a problem asks you to find $\frac{dy}{dx}$ in an equation that cannot be solved for y, you have to resort to the process of *implicit differentiation*, which involves deriving variables with respect to other variables. Whereas in past problems the derivative would be indicated by y' or $f'(x)$, the derivative in implicit differentiation is indicated by $\frac{dy}{dx}$.

> 📖 **DEFINITION**
>
> **Implicit differentiation** allows you to find the slope of a tangent line when the equation in question cannot be solved for y.

Example 3: Find the slope of the tangent line to the graph of $x^2 + 3xy - 2y^2 = -4$ at the point $(1,-1)$.

Solution: Yuck! Clearly this is not solved for y, and if you try to solve for y, you'll get discouraged quickly—solving it for y is impossible due to that blasted y^2. Implicit differentiation to the rescue! The first order of business is finding the derivative of each term of the equation with respect to x. Because you're new at this, I'll go term by term.

The derivative of x^2 with respect to x is $2x$. Nothing fancy is needed, since the variable in the term is the variable we're respecting. However, in the next term, $3xy$, you have to use the Product Rule, since there are two variable terms multiplied ($3x$ and y).

Remember that the derivative of y, with respect to x, is $\frac{dy}{dx}$, so the correct derivative of $3xy$ is $3x \cdot \frac{dy}{dx} + 3 \cdot y$. Finally, the derivative of $-2y^2$ is $-4y \cdot \frac{dy}{dx}$ and the derivative of -4 is 0.

Don't forget to differentiate on *both* sides of the equation! Even though I differentiate implicitly pretty often, I still sometimes forget to differentiate a constant term to get 0. I know; I'm a lunkhead.

All together now, you get a derivative of:

$$2x + 3x \cdot \frac{dy}{dx} + 3y - 4y \cdot \frac{dy}{dx} = 0$$

Move all of the terms not containing a $\frac{dy}{dx}$ to the right side of the equation. Once you've done that, factor the common $\frac{dy}{dx}$ out of the terms on the left side of the equation:

$$3x \cdot \frac{dy}{dx} - 4y \cdot \frac{dy}{dx} = -2x - 3y$$

$$\frac{dy}{dx}\left(3x - 4y\right) = -2x - 3y$$

To finally get the derivative $\left(\frac{dy}{dx}\right)$ by itself, divide both sides of the equation by $3x - 4y$:

$$\frac{dy}{dx} = \frac{-2x-3y}{3x-4y}$$

That's the derivative. The problem asks you to evaluate it at $(1,-1)$, so plug those values in for x and y to get your final answer:

$$\frac{dy}{dx} = \frac{-2(1)-3(-1)}{3(1)-4(-1)}$$

$$= \frac{-2+3}{3+4}$$

$$= \frac{1}{7}$$

YOU'VE GOT PROBLEMS

Problem 2: Find the slope of the tangent line to the graph of $4x + xy - 3y^2 = 6$ at the point (3,2).

Differentiating an Inverse Function

Let's say you're given the function $f(x) = 7x - 5$ and are asked to evaluate $\left(f^{-1}\right)'(1)$, the derivative of the inverse of $f(x)$ when $x = 1$. To find the answer, you would first find the inverse function (using the process we reviewed in Chapter 3) and then find the derivative. However, did you know that you can evaluate the derivative of an inverse function *even if you can't find the inverse function itself?* (Insert dramatic soap opera music here.) You'll learn how to do it in just a second, but we have to review one skill first.

It's important that you're able to find values for an inverse function given only the original function before we try anything more difficult. The procedure we'll use is based on one of the most important properties of inverse functions: if the point (a,b) is on the graph of $f(x)$, then the point (b,a) is on the graph of $f-1(x)$. In other words, if $f(a) = b$, then $f-1(b) = a$.

Example 4: If $g(x) = x^3 + 2$, evaluate $g^{-1}(1)$.

Solution:

Method 1: The easiest way to do this is to figure out exactly what $g^{-1}(1)$ is and then plug in 1. According to our procedure from Chapter 3, here's how you'd go about doing that:

$$y = x^3 + 2$$
$$x = y^3 + 2$$
$$y^3 = x - 2$$
$$y = \sqrt[3]{x - 2}$$
$$g^{-1}(x) = \sqrt[3]{x - 2}$$

Therefore, $g^{-1}(1) = \sqrt[3]{1 - 2} = \sqrt[3]{-1} = -1$. However, there is another way to do this without actually finding $g^{-1}(1)$ first.

Method 2: You're asked to find the *output* of $g^{-1}(x)$ when its input is 1. Remember, I just said that $f(a) = b$ implies $f^{-1}(b) = a$, so therefore the output of $g^{-1}(x)$ when you input 1 is the same exact thing as the input of the original function $g(x)$ when you output 1. So set the original function equal to 1 and solve; the solution will be $g^{-1}(1)$:

$$x^3 + 2 = 1$$
$$x^3 = -1$$
$$x = \sqrt[3]{-1}$$
$$x = -1$$

Either method gives you the same answer.

YOU'VE GOT PROBLEMS

Problem 3: Use the technique of Example 4, Method 2 to evaluate $f^{-1}(6)$ if $f(x) = \sqrt{2x^3 - 18}$.

CRITICAL POINT

Here's a quick summary of this inverse function trick. If I want to evaluate $f^{-1}(a)$, set $f(x) = a$ and solve for x.

Now that you possess this skill, we can graduate to finding values of the derivative of a function's inverse (say that 10 times fast, I dare you). As is the case with just about everything in calculus, there is a theorem governing this practice:

$$\left(f^{-1}\right)'(x) = \frac{1}{f'\left(f^{-1}(x)\right)}$$

So evaluating the derivative is as simple as plugging the value into this slightly more complex, fractiony-looking formula. Once you substitute, your first objective will be to evaluate $f^{-1}(x)$ in the denominator (a skill which we just finished practicing, by no small coincidence).

By the way, are you wondering where this formula comes from? It is pretty easy to generate. Start with the simple inverse function property $f(f^{-1}(x)) = x$ and take the derivative with the Chain Rule:

$$f'\left(f^{-1}(x)\right) \cdot \left(f^{-1}\right)'(x) = 1$$

$$\left(f^{-1}\right)'(x) = \frac{1}{f'\left(f^{-1}(x)\right)}$$

Example 5: If $f(x) = x^3 + 4x + 1$, evaluate $\left(f^{-1}\right)'(2)$.

Solution: According to the formula you learned only moments ago:

$$\left(f^{-1}\right)'(2) = \frac{1}{f'\left(f^{-1}(2)\right)}$$

Start by evaluating $f^{-1}(2)$, which is the equivalent of solving the equation $x^3 + 4x + 1 = 2$. This is not an easy equation to solve; in fact, you can't do it by hand. You'll have to use some form of technology to solve the equation, whether it be a graphing calculator equation solver or a mathematical computer program.

One way to solve this equation is to set the equation equal to zero and calculate the x-intercept on a graphing calculator. See the final section of this chapter ("Technology Focus: Solving Gross Equations") for step-by-step walkthroughs that explain how to solve equations with your graphing calculator (including this gross equation).

Whichever method you choose, the answer is $x = 0.2462661722$, which you can plug into the formula:

$$\left(f^{-1}\right)'(2) = \frac{1}{f'(0.2462661722)}$$

$$= \frac{1}{3(0.2462661722)^2 + 4}$$

$$= \frac{1}{4.1819411}$$

$$\approx 0.239$$

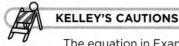

KELLEY'S CAUTIONS

The equation in Example 5 may be difficult to solve, but it is just plain impossible to calculate the inverse function of $f(x) = x^3 + 4x + 1$ using our techniques. So the hard equation is the only way to get an answer at all!

I know that's a lot of decimals, but I didn't want to round any of them until the final answer, or it would have compounded the inaccuracy with every step.

YOU'VE GOT PROBLEMS

Problem 4: If $g(x) = 3x^5 + 4x^3 + 2x + 1$, evaluate $(g^{-1})'(-2)$.

Parametric Derivatives

In order to find a parametric derivative, you differentiate both the x and y components separately and divide the y derivative by the x derivative. In fancy-schmancy mathematical form, it looks like this:

$$\frac{dy}{dx} = \frac{\frac{dy}{dt}}{\frac{dx}{dt}}$$

This formula suggests that you should derive with respect to t, but you should derive with respect to whatever parameter appears in the problem. In the following example, for instance, you'll derive with respect to θ.

Example 6: Find the slope of the tangent line to the parametric curve defined by $x = \cos \theta$ and $y = 2\sin \theta$ when $\theta = \frac{5\pi}{6}$ (pictured in Figure 10.4).

Solution: Since the parameter in these equations is θ, the derivative of the set of parametric equations is:

$$\frac{dy}{dx} = \frac{\frac{dy}{d\theta}}{\frac{dx}{d\theta}}$$

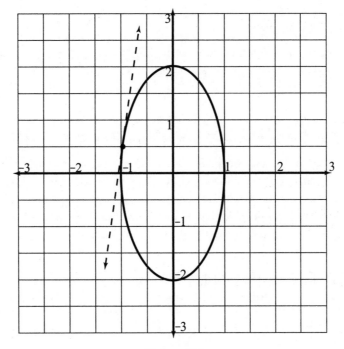

Figure 10.4

The graph of the parametric curve defined by $x = \cos \theta$ *and* $y = 2\sin \theta$ *with the tangent line drawn at* $\theta = \frac{5\pi}{6}$.

Calculate each derivative:

$$\frac{dx}{d\theta} = -\sin\theta \ \text{ and } \ \frac{dy}{d\theta} = 2\cos\theta$$

Finally, calculate the derivative when $\theta = \frac{5\pi}{6}$:

$$\frac{dy}{dx} = \frac{2\cos\frac{5\pi}{6}}{-\sin\frac{5\pi}{6}} = \frac{2\left(-\frac{\sqrt{3}}{2}\right)}{-\left(\frac{1}{2}\right)} = \frac{-\sqrt{3}}{-\left(\frac{1}{2}\right)} = \frac{\sqrt{3}}{\frac{1}{2}}$$

Multiply the numerator by the reciprocal of the denominator to simplify the complex fraction:

$$\frac{dy}{dx} = \sqrt{3} \cdot \frac{2}{1} = 2\sqrt{3}$$

The second derivative (which, like all second derivatives, has the almost incomprehensible notation $\frac{d^2 y}{dx^2}$) of parametric functions is not just the derivative of the first derivative. Instead, it is the derivative of the first derivative divided by the derivative of the original x term:

$$\frac{d^2 y}{dx^2} = \frac{\frac{d}{dt}\left(\frac{dy}{dx}\right)}{\frac{dx}{dt}}$$

> **YOU'VE GOT PROBLEMS**
>
> Problem 5: Determine $\frac{dy}{dx}$ and $\frac{d^2y}{dx^2}$ (the first and second derivatives) for the parametric equations $x = 2t - 3$ and $y = \tan t$.

Technology Focus: Solving Gross Equations

Solving linear equations is a snap once you've had enough practice. Quadratics take a little more work, but with the handy quadratic formula in your nerdy mathematical fanny pack/tool belt, quadratic equations are harmless. However, once you run across equations raised to the third degree or higher, all bets are off. These equations follow no mortal law. It's the Thunderdome and you're Mad Max, but instead of cool face paint and cars, all you have is a graphing calculator and an unsharpened pencil.

In the last section of this chapter, I'll show you how to use your graphing calculator to bring third-degree and higher equations (in other words, *gross equations*) to justice.

Using the Built-In Equation Solver

Both the TI-84 and TI-89 families of calculators have similar equation-solving functionality. Let's look at both as we try to solve the equation $2x^2 - 19x = -35$. To be fair, that's not a completely gross equation because it's a quadratic and because it's actually factorable if you add 35 to both sides and set it equal to 0. The solutions are $x = \frac{5}{2}$ and $x = 7$. Once we practice with this simple equation, we can set our sights on bigger and more dangerous game.

Let's look at the TI-89 first. When you turn the calculator on or press the APPS button, you should see something like Figure 10.5. Select the "Numeric Solver" option.

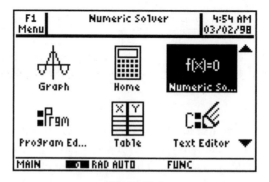

Figure 10.5
Solutions to life's equations are a few button presses away.

Now type your equation into the solver (see Figure 10.6) and press ENTER.

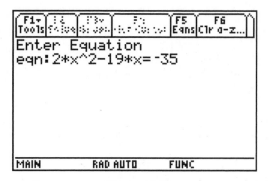

Figure 10.6
The TI-89 has something the TI-84 calculators don't have: an equal sign. You can type your equations verbatim into the solver.

You're prompted for a guess. We know the solutions already, but pretend for a moment that we don't. Let's guess 1, which is close to the actual solution of $x = \frac{5}{2} = 2.5$. Don't worry about the "bound" line beneath your guess—just leave that alone.

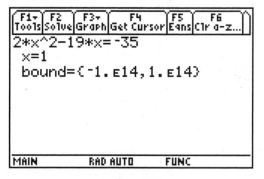

Figure 10.7
Can the calculator solve the equation? The tension in the air is palpable ....

Warning: nothing actually happens if you press enter on your guess. You have to select the "solve" option by pressing [F2]. After muttering to itself for a few moments, the calculator displays its solution, in Figure 10.8.

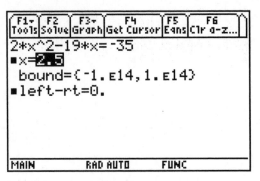

Figure 10.8

Oh, trusty calculator, how could we have ever doubted you?

The process works very similarly with the TI-84. Access the solver by pressing the MATH button and scrolling to "B:Solver…" (see Figure 10.9).

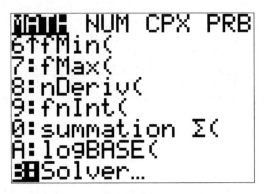

Figure 10.9

The solver is a little harder to find on the TI-84.

There's a *key* difference in the TI-84 solver (pun intended): the "=" key is missing, so *your equation must be set equal to 0*. In order to set the equation $2x^2 - 19x = -35$ equal to 0, you need to add 35 to both sides: $2x^2 - 19x + 35 = 0$. Type that into the solver and press ENTER (see Figure 10.10). (To change the equation once you've typed it in, press the "up" button.)

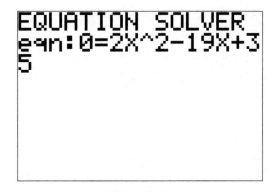

Figure 10.10
Remember, equations must be equal to 0 to use the TI-84 solver.

You are prompted to guess at the answer, just like the TI-89. For grins, let's guess 9, which is close to the actual solution of $x = 7$ (see Figure 10.11).

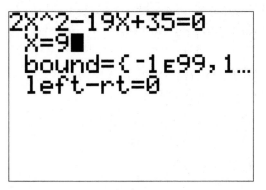

Figure 10.11
There are two solutions to this equation. Your guess determines the solution provided by your calculator.

Again, don't mess around with the bounds settings. The ENTER button doesn't do anything, so make sure to press 2nd – ENTER, which activates the "Solve" button, written in green above the ENTER key. As you might expect, the calculator deftly returns the correct answer of $x = 7$ (see Figure 10.12).

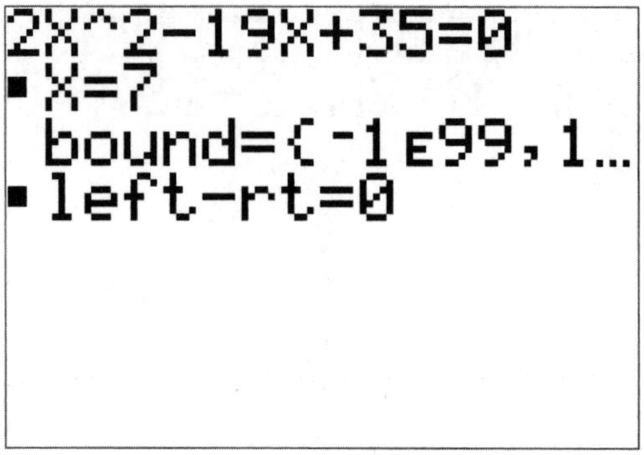

```
2X^2-19X+35=0
■X=7
 bound={-1ε99,1…
■left-rt=0
```

Figure 10.12
The second solution to the equation $2x^2 - 19x = -35$.

The Equation-Function Connection

You may be asking yourself a key question here: how am I supposed to guess a solution? Great question—glad you asked. The easiest way to generate a guess is to look at a graph of the equation. Simply set your equation equal to 0 and then type it into the [Y=] screen.

Let's turn our attention to the gross equation from Example 5: $x^3 + 4x + 1 = 2$. Set this equation equal to 0 by subtracting 2 from both sides:

$$x^3 + 4x + 1 - 2 = 0$$
$$x^3 + 4x - 1 = 0$$

Now enter this equation into the [Y=] screen, as demonstrated in Figure 10.13.

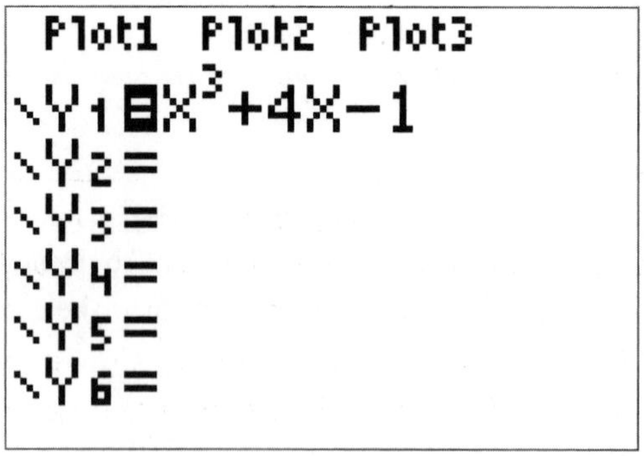

```
Plot1  Plot2  Plot3
\Y₁■X³+4X-1
\Y₂=
\Y₃=
\Y₄=
\Y₅=
\Y₆=
```

Figure 10.13
The Y variable takes the place of the 0 in the equation you set equal to 0. Why?
Keep reading to find out.

The graph of the function (Figure 10.14) crosses the x-axis somewhere between $x = 0$ and $x = 1$. In other words, the equation equals 0 somewhere on the interval $(0,1)$. Because the graph crosses the x-axis only once, you know that the original equation, $x^3 + 4x + 1 = 2$, has only one solution.

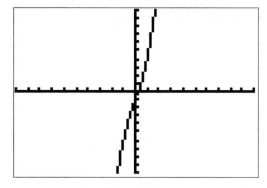

Figure 10.14
The solution to the equation $x^3 + 4x + 1 = 2$ *is also the* x*-intercept of the function*
$Y_1 = x^3 + 4x - 1$.

You can use the solver to calculate the solution to the equation, and 0.5 would be a terrific guess (see Figure 10.15).

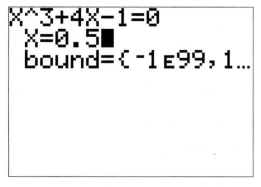

Figure 10.15
Your solver should look like this just before you press the "Solve" button.

The solution is $x \approx 0.2462661722$. Here's the key connection you'll want to remember: a solution to an equation is equal to a root of a function if you create that function by setting the equation equal to 0.

The Least You Need to Know

- To write the equation of a tangent line, use the point of tangency and the derivative at that point in conjunction with the point-slope form of a line.
- You must differentiate implicitly if an equation cannot be solved for y.
- The derivative of a function's inverse is given by the formula $\left(f^{-1}\right)'(x) = \frac{1}{f'\left(f^{-1}(x)\right)}$.
- To calculate a parametric derivative, divide the derivative of the y equation by the derivative of the x equation.
- Solutions to equations are equivalent to roots of functions when you create those functions by setting the equations equal to 0.
- You can use your calculator's built-in solver or x-intercepts to solve gross equations.

CHAPTER 11

Using Derivatives to Graph

Though astrologers have maintained for decades that an individual's astrological sign provides insight into his or her personality, tendencies, and fate, many people remain unconvinced, deeming such thoughts absurd or (in extreme cases) poppycock. (This could be due to the fact that statements such as "The moon is in the third house of Pluto" sound like the title of a new-age Disney movie.) Astrologers don't realize how close they actually came to the truth. It turns out that the signs of the derivatives of a function determine and explain the function's behavior.

In fact, the sign of the first derivative of a function explains what direction that function is heading, and the sign of the second derivative accurately predicts the concavity of the function. It is the third derivative of a function, however, that is able to predict when you will find true love, success in business, and how many times a week it's healthy to eat eggs for breakfast. The easiest way to visualize the signs of a function is via a wiggle graph, which sounds racy but is really quite ordinary when all is said and done.

In This Chapter

- Critical numbers and relative extrema
- Understanding wiggle graphs
- Determining direction and concavity
- The Extreme Value Theorem

Relative Extrema

One common human tendency is to compare oneself with his or her peers on a regular basis. You probably catch yourself doing this all the time, thinking things like, "Of all my friends, I am definitely the funniest." Perhaps you compare more mundane things, like being the best at badminton or having the loudest corduroy pants. However, when you go outside your social sphere, you often find someone who is significantly funnier than you or who possesses supersonically loud pants. This illustrates the difference between a relative extreme point and an absolute extreme point. You can be the smartest of a group of people without being the smartest person in the world.

For example, look at the graph in Figure 11.1, with points of interest *A, B, C, D,* and *E* noted.

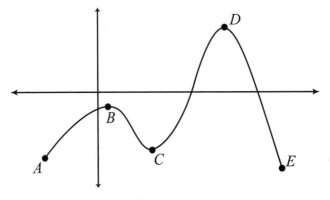

Figure 11.1
This graph has only one absolute maximum and one absolute minimum, but several relative extrema.

The absolute maximum on the graph occurs at *D,* and the absolute minimum on the graph occurs at *E.* However, the graph has a *relative* maximum at *B,* and a *relative* minimum at *C.* These may not be the highest or lowest points of the entire graph, but (as little hills and valleys) they are the highest and lowest points in their immediate vicinity.

A relative extrema point (whether a maximum or a minimum) occurs when that point is higher or lower than all of the points around it. Visually, a relative maximum is the peak of a hill in the graph, and a relative minimum is the lowest point of a dip in the graph. Absolute extrema points are the highest or lowest of all the relative extrema on a graph. Remember that the term *extrema* is just plural for "extremely high or low point."

Finding Critical Numbers

A *critical number* is an x value that causes a function either to equal zero or become undefined. They're extremely useful for finding extrema points because a function, $f(x)$, can only change direction at a critical number of its derivative, $f'(x)$. Why? When $f'(x)$ is 0, then $f(x)$ is neither increasing ($f'(x) \geq 0$) nor decreasing ($f'(x) \leq 0$), meaning $f(x)$ is most likely about to do something drastic.

> **DEFINITION**
>
> A **critical number** is an x value that either makes a function zero or undefined.

Example 1: Given $f(x) = x^3 - x^2 - x + 2$, identify $f'(x)$ and its critical numbers.

Solution: Begin by finding the derivative of $f(x)$, then set it equal to 0 and solve:

$$f'\left(x\right) = 3x^2 - 2x - 1$$
$$0 = 3x^2 - 2x - 1$$
$$0 = \left(3x + 1\right)\left(x - 1\right)$$
$$x = -\tfrac{1}{3}, 1$$

There are no places where $f'(x)$ does not exist, so $x = -\frac{1}{3}$ and $x = 1$ are the only two critical numbers.

If you take a look at the graph of $f(x)$, you'll notice that the graph does, indeed, change direction at those x values (see Figure 11.2).

However, you don't need to use the graph of a function to determine if the graph changes direction, or if it does, whether it causes a relative maximum or minimum.

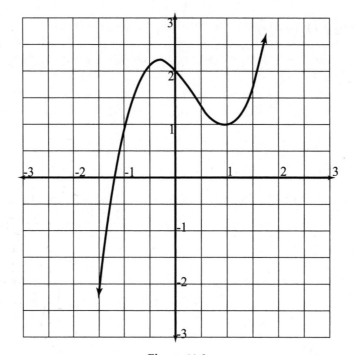

Figure 11.2

This graph changes from increasing to decreasing at $x = -\frac{1}{3}$ *and then returns to increasing once*
$x = 1.$

Classifying Extrema

As I alluded to earlier, the sign of $f'(x)$ tells you whether $f(x)$ is increasing or decreasing. This is true because an increasing graph will have a tangent line with a positive slope and a decreasing graph will possess a negatively sloped tangent line. Therefore, you can tell what's happening between the critical numbers of $f'(x)$ (i.e., if $f(x)$ is increasing or decreasing) by picking some points on the graph between the critical numbers and determining whether the derivatives there are positive or negative.

Example 2: If $f(x) = x^3 - x^2 - x + 2$ and the critical numbers of its derivative, $f'(x)$, are $x = -\frac{1}{3}$ and $x = 1$, describe the direction of $f(x)$ between those critical numbers using the sign of $f'(x)$.

Solution: Choose three x values, one less than the first critical number, one between the critical numbers, and one greater than the second. I will choose simple values to make my life easier: $x = -1$, 0, and 2. Plug each of these x's into $f'(x)$, and the sign of the result will tell you if the function $f(x)$ is increasing or decreasing there:

$$f'(-1) = 3(-1)^2 - 2(-1) - 1 \quad = 3 + 2 - 1 \quad = 4$$
$$f'(0) = 3(0)^2 - 2(0) - 1 \quad\quad = 0 - 0 - 1 \quad = -1$$
$$f'(2) = 3(2)^2 - 2(2) - 1 \quad\quad = 12 - 4 - 1 = 7$$

Because $f'(x)$ is positive when $x = -1$, and $x = -1$ comes before the first critical point, the function will be increasing until $x = -\frac{1}{3}$, the first critical number. However, the derivative turns negative between the critical numbers, so $f(x)$ is decreasing between $x = -\frac{1}{3}$ and 1.

After $x = 1$, the derivative turns positive again, so $f(x)$ will increase beyond that point. This is no giant surprise, because you already saw the graph of $f(x)$ in Figure 11.2, but notice how the critical numbers create regions where the graph is going different directions (up and down) as you travel along the graph from left to right. Figure 11.3 shows the signs of $f'(x)$ and the graph of $f(x)$ at the same time to help you visualize what's going on.

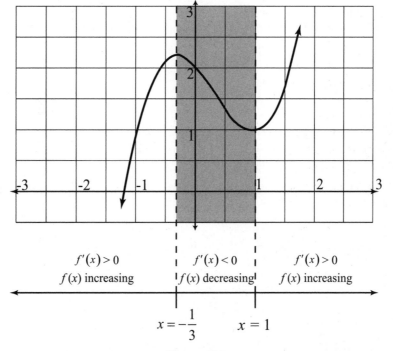

Figure 11.3

Notice how the sign of the derivative correlates with the direction of the original function. The number line is actually a wiggle graph, as you'll learn in the next section.

YOU'VE GOT PROBLEMS

Problem 1: Given $h(x) = -x^2 + 6x + 27$, calculate the critical number of $h'(x)$, and determine whether or not it represents a relative maximum or minimum, based on the signs of $h'(x)$.

The Wiggle Graph

A *wiggle graph* is a nice, compact way to visualize the signs of a function's derivative all at once. To create a wiggle graph, we'll use the procedure from Example 2. In other words, we'll find the critical numbers, pick "test values" between those critical numbers, and plug those into the derivative to determine the direction of the function. The result will be a number line, segmented by critical numbers, and labeled with the signs of the derivative in each of its intervals. This will help us to quickly find all relative extreme points on the graph.

DEFINITION

A **wiggle graph** (or sign graph) is a segmented number line that describes the direction of a function. It's called a wiggle graph because it tells you which way the graph is wiggling (i.e., if it is increasing or decreasing).

Example 3: Create a wiggle graph for the function $f(x) = \frac{x^2+2x+1}{x-5}$ and use it to determine which critical numbers are relative extrema.

Solution: First you must find the critical numbers, where $f'(x)$ is either equal to 0 or is undefined. You use the Quotient Rule to find $f'(x)$:

$$f'(x) = \frac{(x-5)(2x+2) - (x^2+2x+1)}{(x-5)^2}$$

$$= \frac{x^2 - 10x - 11}{(x-5)^2}$$

Because this is a fraction, it is equal to 0 when the numerator equals 0 and undefined when the denominator is equal to 0. Both of these events interest us for the purpose of finding critical numbers, so factor the numerator and set both it and the denominator equal to 0 and solve:

$$f'(x) = \frac{(x-11)(x+1)}{(x-5)^2}$$

The derivative equals 0 when $x = 11$ or $x = -1$ and is undefined when $x = 5$, so these are the critical numbers. Draw a number line and mark these numbers on it as shown in Figure 11.4.

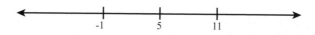

Figure 11.4

The beginnings of a wiggle graph. Note that you don't need to label any numbers on it except the critical numbers; you don't even have to worry about drawing it to scale. It's just a tool for visualization, not a scientific graph.

These three critical numbers split the number line into four intervals. Remember that the function will always go in the same direction during each interval, because it can only change direction at a critical number. Therefore, you can choose any number in each interval as a "test value." I'll choose the numbers $x = -2$, 0, 6, and 12. Now, plug these four numbers into the derivative:

$$f'(-2) = \frac{(-13)(-1)}{(-7)^2} \qquad = \frac{13}{49}$$

$$f'(0) \ = \frac{(-11)(1)}{(-5)^2} \qquad = -\frac{11}{25}$$

$$f'(6) \ = \frac{(-5)(7)}{(1)^2} \qquad = -35$$

$$f'(12) \ = \frac{(1)(13)}{(7)^2} \qquad = \frac{13}{49}$$

Because $f'(-2)$ is positive, $f'(x)$ is actually positive for the entire interval $(-\infty,-1)$, so indicate that with a "+" above the interval in the wiggle graph. Similarly, the first derivative is negative on the interval $(-1,5)$, negative on $(5,11)$, and positive on $(11,\infty)$. In Figure 11.5, this information is summed up on the wiggle graph.

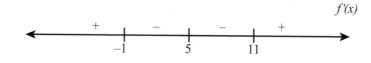

Figure 11.5

The signs of f '(x) correspond to the direction of f '(x). Positive means increasing, negative means decreasing. Note that the wiggle graph is clearly labeled "f '(x)." Always label your wiggles to avoid confusion.

Now you can tell that the function changes direction from increasing to decreasing at $x = -1$. If you plug that critical number into $f(x)$, you get the critical and relative maximum point $(-1,0)$. Similarly, a sign change at $x = 11$ indicates a relative minimum at the critical point $(11,24)$.

> **YOU'VE GOT PROBLEMS**
>
> Problem 2: Draw the wiggle graph for the function
> $g(x) = 2x^3 - \frac{7}{2}x^2 - 3x + 10$, and determine the intervals on which $g(x)$ is increasing.

The Extreme Value Theorem

Your first experience with existence theorems was the Intermediate Value Theorem (see Chapter 7). Do you remember it fondly? I guess that's a rhetorical question, because whether you liked it or not, here comes your second existence theorem. The Extreme Value Theorem, like its predecessor, really doesn't say anything earth shattering, but it should make a lot of sense, so that's a plus.

The Extreme Value Theorem: If a function $f(x)$ is continuous on the closed interval $[a,b]$, then $f(x)$ has an absolute maximum and an absolute minimum on $[a,b]$.

> **KELLEY'S CAUTIONS**
>
> Before you conclude that a sign change in a wiggle graph indicates a relative extrema point, make sure that the original function is defined there! For example, in $f(x) = \frac{1}{x^2}$, the function changes from increasing to decreasing at $x = 0$ (verify with a wiggle graph of $f'(x)$). However, $x = 0$ is not in the domain of $f(x)$, so it cannot be a relative maximum.

This theorem simply says that a piece of continuous function will always have a highest point and a lowest point. That's all. Here's a little tip: a function's absolute extrema can only occur at one of two places—either at a *relative* extrema point or at an endpoint. This little trick makes finding the absolute extrema points very easy.

Example 4: Find the absolute maximum and absolute minimum of the function
$f(x) = \frac{3}{5}x^5 - \frac{2}{3}x^3 - x + 2$ on the interval $[-2,1]$.

Solution: The absolute extrema you're looking for are guaranteed to exist according to the Extreme Value Theorem, because $f(x)$ is continuous on the closed interval. In fact,

$f(x)$ is continuous everywhere! Start by drawing a wiggle graph. Same process as always: set $f'(x)$ = 0 and plug test values into the derivative:

$$f'(x) = 3x^4 - 2x^2 - 1$$
$$0 = (3x^2 + 1)(x^2 - 1)$$
$$0 = (3x^2 + 1)(x + 1)(x - 1)$$
$$x = -1, 1$$

Check out the wiggle graph in Figure 11.6. Because the sign of its derivative changes at both critical numbers (and they are both in the domain of $f(x)$), you know that $x = -1$ and 1 mark relative extrema and, therefore, possibly absolute extrema as well.

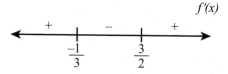

Figure 11.6

According to this wiggle graph, f(x) changes direction twice.

Because an extreme value (an absolute maximum or minimum) can only occur at a critical number ($x = -1$ or 1) or an endpoint ($x = -2$ or 1), plug each of those x values into $f(x)$ to see which yields the highest and lowest values:

$$f(-2) = -\frac{148}{15} \approx -9.867$$
$$f(-1) = \frac{46}{15} \approx 3.067$$
$$f(1) = \frac{14}{15} \approx 0.9333$$

KELLEY'S CAUTIONS

There is no solution to the mini-equation $3x^2 + 1 = 0$ in Example 4, because solving it gives you $x = \pm\sqrt{-\frac{1}{3}}$, and you can't take the square root of a negative number.

Therefore, the absolute maximum of $f(x)$ on the closed interval $[-2,1]$ will be $\frac{46}{15}$ and the absolute minimum is $-\frac{148}{15}$. I know that those fractions were ugly, but whatever doesn't kill you makes you stronger, right? You're not buying that, are you?

KELLEY'S CAUTIONS

Reporting an absolute maximum of –1 and an absolute minimum of –2 for Example 4 is a common error. Although these *are* the *x* values where the extrema occur, they are not the extreme values themselves. Absolute maxima and minima are *heights*—function values, not *x* values.

YOU'VE GOT PROBLEMS

Problem 3: Find the absolute maximum and minimum of $g(x) = x^3 + 4x^2 + 5x - 2$ on the closed interval [–5,2].

Determining Concavity

Concavity describes how a curve bends. A curve that can hold water poured into it from the top of the graph is said to be "concave up," whereas one that cannot hold water is said to be "concave down." Notice that the concave-up curve in Figure 11.7 would catch water poured into it, whereas the concave-down curve would dump the water onto the floor, causing your mother to get angry.

DEFINITION

The **concavity** of a curve describes the way the curve bends. A curve that is "concave up" would catch water, whereas one that is "concave down" would dump the water.

The sign of the second derivative $f''(x)$ describes the concavity of $f(x)$. If $f''(x)$ is positive for some *x* value, then $f(x)$ is concave up at that point. If, however, $f''(x)$ is negative, then $f(x)$ is concave down at that point. You can remember this relationship between the second derivative's sign and concavity using Figure 11.8.

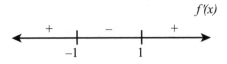

$f'(x)$

Figure 11.7
A tale of two curves whose second derivatives differ. (You'll soon see what I mean.)

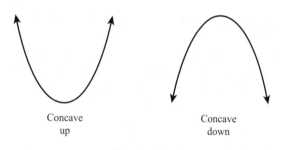

Figure 11.8
A smile is concave up, indicating a positive second derivative via the plus-sign eyes. You'd be unhappy too, if you were concave down.

The sign of $f''(x)$ not only describes the concavity of $f(x)$, it also describes the direction of $f'(x)$. This is because $f''(x)$ is also the first derivative of $f'(x)$, and remember that first derivatives describe the direction of their predecessors. For example, if $g''(2) = -7$ for some function $g(x)$, then we know $g(x)$ is concave down when $x = 2$ (because the second derivative is negative) *and* we know $g'(x)$ is *decreasing* at $x = 2$.

As with direction, however, the concavity of a curve can change throughout the function's domain (the points of change are called *inflection points*). You'll use a process that mirrors the first-derivative wiggle graph to determine a function's concavity.

> **DEFINITION**
>
> A graph changes concavity at an **inflection point.**

Another Wiggle Graph

Hopefully you've seen how useful a wiggle graph can be in helping you visualize a function's direction. It is just as useful when visualizing concavity, and is just as easy. This time, you'll use the second derivative to create the wiggly number line, and you'll plug test values into the second derivative rather than the first derivative to come up with the appropriate signs. Let's revisit an old friend, $f(x)$, from Example 4.

Example 5: On what intervals is the function $f(x) = \frac{3}{5}x^5 - \frac{2}{3}x^3 - x + 2$ concave up?

Solution: Find the second derivative, $f''(x)$, and use it to create a wiggle graph, as you did earlier in the chapter. The only difference is that you'll use $f''(x)$ for everything instead of $f'(x)$:

$$f'(x) = 3x^4 - 2x^2 - 1$$
$$f''(x) = 12x^3 - 4x$$

Set $f''(x) = 0$ and solve for x to get your critical numbers:

$$4x(3x^2 - 1) = 0$$
$$x = 0, \pm\sqrt{\tfrac{1}{3}}$$

Don't forget the $\pm$ sign, because you are square-rooting both sides of an equation. It's time to draw the wiggle graph and choose test points just like before. Because you already know how to do this, let's jump straight to the correct graph in Figure 11.9.

Figure 11.9
The second derivative wiggle graph for f(x).

The function $f(x)$ is concave up whenever $f''(x)$ is positive, so $f(x)$ is concave up on $\left(-\sqrt{\tfrac{1}{3}}, 0\right)$ and $\left(\sqrt{\tfrac{1}{3}}, \infty\right)$.

> **YOU'VE GOT PROBLEMS**
>
> Problem 4: When is $f(x) = \cos x$ concave down on $(0, 2\pi)$?

The Second Derivative Test

The Second Derivative Test is a little math trick that tells you whether or not an extrema point is a relative maximum or minimum. You did this using the signs of the first derivative and a wiggle graph earlier. However, the Second Derivative Test uses the sign of the second derivative (and therefore the concavity of the graph at that point) to do all the work.

The Second Derivative Test: Plug the critical numbers that occur when $f'(x) = 0$ or $f'(x)$ is undefined into $f''(x)$. If the result is positive, that critical number is a relative minimum on $f(x)$. If the result is negative, that critical number marks a relative maximum on $f(x)$. If the result is 0, you cannot draw any conclusion from the Second Derivative Test and must resort to the first-derivative wiggle graph.

Example 6: Classify all the relative extrema of the function $g(x) = 3x^3 - 18x + 1$ using the Second Derivative Test.

Solution: First find the critical numbers as you did earlier in the chapter:

$$g'(x) = 9x^2 - 18$$
$$0 = 9x^2 - 18$$
$$18 = 9x^2$$
$$2 = x^2$$
$$\pm\sqrt{2} = x$$

CRITICAL POINT

If you think about it, the only possible extrema point you can have on a concave-up graph is a relative minimum—consider the point (0,0) on the graph of $y = x^2$ as an example.

Plug both $x = \sqrt{2}$ and $x = -\sqrt{2}$ into $g''(x) = 18x$. Because $g''\left(\sqrt{2}\right) = 18\sqrt{2}$, which is positive, $x = \sqrt{2}$ represents the location of a relative minimum (according to the Second Derivative Test). Conversely, because $g''\left(-\sqrt{2}\right) = -18\sqrt{2}$, that point represents a relative maximum.

The Least You Need to Know

- Critical numbers are x values that cause a function to equal 0 or become undefined. The graph of $f(x)$ can only change direction at a critical number of its derivative, $f'(x)$.

- If $f'(x)$ is positive, then $f(x)$ is increasing; a negative $f'(x)$ indicates a decreasing $f(x)$.

- If $f''(x)$ is positive, then $f(x)$ is concave up; a negative $f''(x)$ indicates a concave-down $f(x)$.

- You can use both the first derivative wiggle graph and the Second Derivative Test to classify relative extrema points.

CHAPTER 12

Derivatives and Motion

Mathematics can actually be applied in the real world. This may shock and appall you, but it's true. It's probably shocking because most of the problems we've dealt with have been purely computational in nature, devoid of correlation to real life. (For example, estimating gas mileage is a useful mathematical real-life skill, whereas factoring difference of perfect cube polynomials is not as useful in a straightforward way.) Most people hate real-life application problems because they are (insert scary wolf howl here) *word problems!*

Factoring and equation solving may be rote, repetitive, and a little boring, but at least they're predictable. How many nights have you gone to sleep haunted by problems like this: "If Train *A* is going from Pittsburgh to Los Angeles at a rate of 110 kilometers per hour, Train *B* is traveling 30 kilometers less than half the number of male passengers in Train *A*, and the heading of Train *B* is 3 degrees less than the difference of the prices of a club sandwich on each train, then at what time will the conductor of the first train remember that he forgot to set the DVR to record *Jeopardy?*"

Position equations are a nice transition into calculus word problems. Even though they are slightly bizarre, they follow clear patterns. Furthermore, they give you the chance to show off your new derivative skills.

In This Chapter

* What is a position equation?
* The relationship between position, velocity, and acceleration
* Speed versus velocity
* Understanding projectile motion

The Position Equation

A *position equation* is an equation that mathematically models something in real life. Specifically, it gives the position of an object at a specified time. Different books and teachers use different notation, but I will always indicate a position equation with the notation $s(t)$, to stay consistent. By plugging values of t into the equation, you can determine where the object in question was at that moment in time. Just in case you're starting to get stressed out, I'll insert something cute and cuddly into the mix—my cat, Peanut.

> **DEFINITION**
>
> A **position equation** is a mathematical model that outputs an object's position at a given time t. Position is usually given with relation to some fixed landmark, like the ground or the origin, so that a negative position means something. For example, $s(5) = -6$ may mean that the object in question is 6 feet below the origin after 5 seconds have passed.

Peanut pretty much has the run of my basement, and her favorite pastime (apart from her strange habit of chewing on my eyeglasses) is batting a ball back and forth along one of the basement walls. For the sake of ease, let's say the wall in question is 20 feet long; we'll call the exact middle of the wall position 0, the left edge of the wall (our left, not her left) position −10, and the right edge of the wall position 10, as in Figure 12.1.

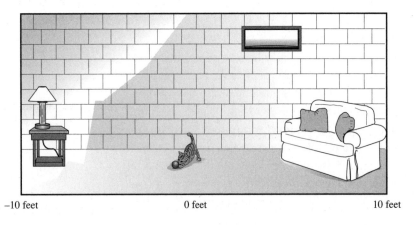

−10 feet   0 feet   10 feet

Figure 12.1
The domain of Peanut the cat. For sake of reference, I have labeled the middle and edges of the room.

Let's examine the kitty's position versus time in a simple example. We'll keep returning to this example throughout the chapter as we compound our knowledge of derivatives and motion (and cat recreation).

Example 1: During the first four seconds of a particularly frisky playtime, Peanut's position (in feet at time = t seconds) along the wall is given by the equation $s(t) = t^3 - 3t^2 - 2t + 1$. Evaluate and explain what is meant by $s(0)$, $s(2)$, and $s(4)$.

Solution: Plug each number into $s(t)$. A positive answer means she is toward the right of the room, whereas a negative answer means she is to the left of center. The larger the number, the farther she is to the right or left:

$$s(0) = 0^3 - 3(0)^2 - 2(0) + 1 = 1$$

$$s(2) = 2^3 - 3(2)^2 - 2(2) + 1 = 8 - 12 - 4 + 1 \quad = -7$$

$$s(4) = 4^3 - 3(4)^2 - 2(4) + 1 = 64 - 48 - 8 + 1 \quad = 9$$

 KELLEY'S CAUTIONS

The position given by $s(t)$ in Example 1 is the horizontal position of the cat—had it meant vertical position, that negative answer would have been disturbing. Every position problem should infer what is meant by its output and will usually include units (such as feet and seconds) as well.

Therefore, when $t = 0$ (i.e., before you start measuring elapsed time), $s(0) = 1$ tells you the cat began 1 foot to the right of the center. Two seconds later ($t = 2$), she had used her lightning-fast kitty movements to travel 8 feet left, meaning she was then only 3 feet from the left wall, and 7 feet left of center. Two seconds after that ($t = 4$), she had moved 16 feet right, now only 1 foot away from the right-hand wall. That is one fast-moving cat.

CRITICAL POINT

The value $s(0)$ is often called *initial position,* because it gives the position of the object before you start measuring time. Similarly, $v(0)$ and $a(0)$ are the *initial velocity* and *initial acceleration.*

There's nothing really fancy about the position equation; given a time input, it tells you where the object was at that moment. Notice that the position equation in Example 1 is a nice, continuous, and differentiable polynomial. You can find the derivative awfully easily, but what does the derivative of the position equation represent?

> **YOU'VE GOT PROBLEMS**
>
> Problem 1: A particle moves vertically (in inches) along the y-axis according to the position equation $s(t) = \frac{1}{2}t^3 - 5t^2 + 3t + 6$, where t represents seconds. At what time(s) is the particle 30 inches below the origin?

Velocity

Remember that the derivative describes the rate of change of a function. Therefore, $s'(t)$ describes the velocity of the object in question at any given instant. It makes sense that velocity is equivalent to the rate of change of position, because velocity measures how quickly you move from one position to another. *Speed* also measures how quickly something moves, but speed and velocity are not the same thing.

> **CRITICAL POINT**
>
> If an object is moving downward at a rate of 15 feet per second, you could say that its velocity is –15 feet/second, whereas its speed is 15 feet/second. *Speed is always the absolute value of velocity.*

Velocity combines an object's speed with its direction, whereas speed just gives you the rate at which the object is traveling. Practically speaking, this means that velocity can be negative, but speed cannot. What does a negative velocity mean? It depends on the problem. In a horizontal motion problem (like the Peanut the cat problem), it means velocity towards the left (because the left was defined as the negative direction). In a vertical motion problem, a negative velocity typically means that the object is dropping.

To find the velocity of an object at any instant, calculate the derivative and plug in the desired time for t. If, however, you want an object's average velocity (i.e., average rate of change), remember that this value comes from the slope of the secant line. Remember how quickly Peanut was darting around in Example 1? Let's get those exact speeds and velocities using the derivative.

Example 2: Peanut the cat's position, in feet, for $0 \le t \le 4$ seconds is given by $s(t) = t^3 - 3t^2 - 2t + 1$. Find her velocity and speed at times $t = 1$ and $t = 3.5$ seconds, and give her average velocity over the t interval $[1, 3.5]$.

Solution: There are lots of parts to this problem, but none are hard. Start by calculating her velocity at the given times. Remember that the velocity is the first derivative of the position equation, so $s'(t) = v(t) = 3t^2 - 6t - 2$:

$$v(1) \quad = 3 - 6 - 2 \qquad = -5 \text{ ft/sec}$$
$$v(3.5) = 36.75 - 21 - 2 = 13.75 \text{ ft/sec}$$

Peanut is moving at a speed of 5 feet/second to the left (because the velocity is negative) at $t = 1$ second, and she is moving much faster, at a speed of 13.75 feet/second to the right, when $t = 3.5$ seconds. Now to find the average velocity on [1,3.5]—it's equal to the slope of the line segment connecting the points on the position graph where $t = 1$ and $t = 3.5$. To find these points, plug those values into $s(t)$:

$$s(1) = 1 - 3 - 2 + 1 = -3$$
$$s(3.5) = 42.875 - 36.75 - 7 + 1 = 0.125$$

KELLEY'S CAUTIONS

The slope of a position equation's tangent line equals the instantaneous velocity at the point of tangency. The slope of a position equation's secant line gives the average velocity over that interval. Notice that instantaneous and average rates of change are both based on linear slopes drawn on the *position equation,* not its derivative.

Calculate the secant slope using the points $(1,-3)$ and $(3.5, 0.125)$:

$$m = \frac{0.125 - (-3)}{3.5 - 1} = 1.25 \text{ ft/sec}$$

Therefore, even though she runs left and right at varying speeds over the time interval [1,3.5], she averages a rightward speed of 1.25 feet/second.

YOU'VE GOT PROBLEMS

Problem 2: A particle moves vertically (in inches) along the y-axis according to the position equation $s(t) = \frac{1}{2}t^3 - 5t^2 + 3t + 6$, where t represents seconds. Rank the following from least to greatest: the speed when $t = 3$, the velocity when $t = 7$, and the average velocity on the interval [2,6].

Acceleration

As velocity is to position, so is acceleration to velocity. In other words, acceleration is the rate of change of velocity. Think about it—if you're driving in a car that suddenly speeds up, the sense of being pushed back in your seat is due to the effects of acceleration. It is not the high rate of speed that makes roller coasters so scary. Aside from their height, it is the sudden acceleration and deceleration of the rides that causes the passengers to experience dizzying effects (and occasionally their previous meal).

CRITICAL POINT

The units for acceleration will be the same as the units for velocity, except the denominator will be squared. For example, if velocity is measured in feet per second (ft/sec), then acceleration is measured in feet per second per second, or ft/sec².

To calculate the acceleration of an object, evaluate the second derivative of the position equation (or the first derivative of velocity). To calculate average acceleration, find the slope of the secant line on the velocity function (for the same reasons that average velocity is the secant slope on the position function). Let's head back to the cat of mathematical mysteries one last time.

CRITICAL POINT

If the first derivative of position represents velocity and the second derivative represents acceleration, the third derivative represents "jerk," the rate of change of acceleration. Think of jerk as that feeling you get as you switch gears in your car and the acceleration changes. I've never seen a problem concerning jerk, but I have known a few mathematicians who were pretty jerky.

Example 3: Peanut the cat's position, in feet, at any time $0 \leq t \leq 4$ seconds is given by $s(t) = t^3 - 3t^2 - 2t + 1$. When, on the interval $[0,10]$, is she decelerating?

Solution: Because the sign of the second derivative determines acceleration, you want to know when $s''(x)$ is negative. So make an $s''(x)$ wiggle graph by setting it equal to 0, finding critical numbers, and picking test points (as you did in Chapter 11). The wiggle graph for the second derivative is given in Figure 12.2.

$$s'(t) = v(t) = 3t^2 - 6t - 2$$
$$s''(t) = v'(t) = a(t) = 6t - 6$$

$$6t - 6 = 0$$
$$6t = 6$$
$$t = 1$$

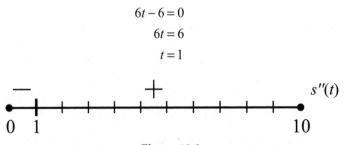

Figure 12.2

Because s″(x) is negative on (0,1), the cat is decelerating on that interval.

The acceleration equation $s''(t)$ is negative on the interval $(0,1)$, so the cat decelerates only between $t = 0$ and $t = 1$.

YOU'VE GOT PROBLEMS

Problem 3: A particle moves vertically (in inches) along the y-axis according to the position equation $s(t) = \frac{1}{2}t^3 - 5t^2 + 3t + 6$, where t represents seconds. At what time t is the acceleration of the particle equal to -1 in/sec^2?

Vertical Projectile Motion

One of the easiest types of motion to model in elementary calculus is projectile motion, the motion of an object acted upon solely by gravity. Have you ever noticed that any thrown object follows a vertical path to the ground? It is very easy to write the position equation describing that path with only a tiny bit of information. Mind you, these equations can't give you the exact position, because ignoring wind resistance and drag makes the problem much easier.

Scientists often pooh-pooh these little pseudoscientific math applications, saying that ignoring such factors as wind resistance and drag renders these examples worthless. Math people usually contend that, although not perfect, these examples show how useful even a simple mathematical concept can be.

CRITICAL POINT

Unquestionably, one of the grossest examples of projectile motion is in the movie *The Exorcist*. We will not be doing any examples involving pea soup.

The position equation of a vertical projectile looks like this:

$$s(t) = \frac{1}{2}g \cdot t^2 + v_0 \cdot t + h_0$$

You plug the object's initial velocity into v_0, the initial height into h_0, and the appropriate gravitational constant into g (which stands for acceleration due to gravity)—if you are working in feet, use $g = -32$, whereas $g = -9.8$ if the problem contains meters. Once you create your position equation by plugging into the formula, it will work just like the other position equations from this chapter, outputting the vertical height of the object in relation to the ground at any time t (for example, a position of 12 translates to a position of 12 feet above ground).

Example 4: Here's a throwback to 1970s television for you. A radio station called WKRP in Cincinnati is running a radio promotion. For Thanksgiving, they are dropping live turkeys from the station's traffic helicopter into the city below, but little do they know that turkeys are not so good at the whole flying thing. Assuming that the turkeys were tossed with a miniscule initial velocity of 2 ft/sec straight up from a safe hovering height of 1,000 feet above ground, how long does it take a turkey to hit the road below, and at what speed will the turkey be traveling at that time?

Solution: The problem contains feet, so use $g = 32$ ft/sec^2; you're given $v_0 = 2$ and $h_0 = 1,000$, so plug these into the formula to get the position equation of $s(t) = -16t^2 + 2t + 1,000$. You want to know when they hit the ground, which means they have a position of 0, so solve the equation $-16t^2 + 2t + 1,000 = 0$. You can use a calculator or the quadratic formula to come up with the answer of $t = 7.9684412$ seconds. (The other answer of -7.84 doesn't make sense—a negative answer suggests going back in time, and that's never a good idea, especially with poultry.) If you plug that value into $s'(x)$, you'll find that the turkeys were falling at a velocity of -252.990 ft/sec. Oh, the humanity.

CRITICAL POINT

Notice that g always equals either -32 or -9.8, depending on the units of the problem. This is because the pull of gravity never changes.

YOU'VE GOT PROBLEMS

Problem 4: A cannonball is fired straight up from a fortification 75 meters above ground with an initial velocity of 100 meters/second. Given this information, answer the following questions:

    (a) When will the cannonball reach its maximum height? Round your answer to the nearest thousandth of a second.

    (b) What is the maximum height of the cannonball? Round your answer to the nearest meter.

    (c) Assuming the cannon is firing into flat terrain, how long does it take the cannonball to first hit the ground?

The Least You Need to Know

- The position equation tells you where an object is at any time t.
- The derivative of the position equation is the velocity equation, and the derivative of velocity is acceleration.
- If you plug 0 into position, velocity, or acceleration, you'll get the initial value for that function.
- The formula for the position of a vertical projectile is $s(t) = \frac{1}{2}g \cdot t^2 + v_0 \cdot t + h_0$.

Common Derivative Applications

It's been a fun ride, but our time with the derivative is almost through. Don't get too emotional yet—I've saved the best for last, and this chapter will be a hoot (if you like word problems, that is). As in the last chapter, we'll be looking at the relationship between calculus and the real world, and you'll probably be surprised by what you can do with very simple calculus procedures.

This chapter has it all: cool shortcuts, a few more existence theorems, romance, adventure, and the two topics most first-year calculus students find the trickiest. We'll go through the topics in the order of difficulty, starting with the easiest and progressing to the more advanced.

In This Chapter

* Approximating zeroes of functions
* Limits of indeterminate expressions
* The Mean Value Theorem
* Rolle's Theorem
* Calculating related rates
* Maximizing and minimizing functions

Newton's Method

You may remember that Sir Isaac Newton was one of the two men responsible for discovering/inventing calculus. One technique named for him allows you to approximate difficult-to-find roots (or zeroes) of a function.

This technique is interesting mostly because of its historic significance rather than its modern-day usefulness. Back in Chapter 10 you learned how to use your calculator to approximate difficult roots, rather than churning them out by hand.

You may be interested to know that your calculator uses a technique very similar to Newton's Method to calculate those roots, and after spending a few moments working with it, you may gain an appreciation for all the work your calculator is doing behind the scenes.

Newton's Method is an *iterative* formula that can be repeated over and over, each time producing a value that is slightly closer to the correct answer (if everything is working correctly). You begin with a seed value, something relatively close to the right answer (like the guesses your calculator required in Chapter 10). Often, that seed value is called x_0, and after running it through the formula you end up with a better guess called x_1. You can then run x_1 through the formula to get an even better guess called x_2 and so on.

> 📖 **DEFINITION**
>
> An **iterative** formula is used repeatedly to sleuth out a specific value. Each time you get a value from the formula, you plug that value back in to get a more accurate value for the next round.

With this in mind, Newton's Method calls your current guess x_n and the resulting better guess x_{n+1}. To generate that better guess you apply this formula:

$$x_{n+1} = x_n - \frac{f(x_n)}{f'(x_n)}$$

In other words, from your original guess x_n, you subtract the function evaluated at x_n divided by the derivative of the function evaluated at x_n.

Example 1: Calculate two iterations of Newton's Method to approximate the positive root of $f(x) = x^2 - 5$ using an initial (seed) value of $x_0 = 2$.

Solution: Look at the graph of $f(x)$ in Figure 13.1. It appears to cross the x-axis at a value slightly greater than 2 and a value slightly less than −2. The problem asks you to estimate the positive root, so the seed value of $x_0 = 2$ seems like a good guess to begin with.

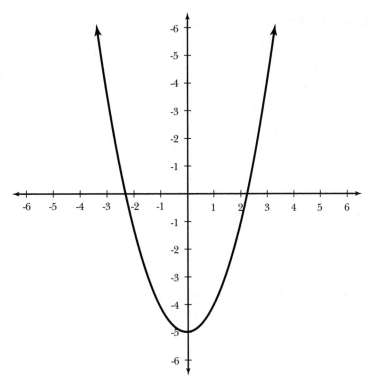

Figure 13.1
The graph of f(x) = x² − 5.

Begin by evaluating the function and its derivative, $f'(x) = 2x$, at the seed value, $x_0 = 2$:

$$f(2) = 2^2 - 5 = -1 \qquad\qquad f'(2) = 2 \cdot 2 = 4$$

Substitute these values into Newton's Method to calculate $x_{n+1} = x_{0+1} = x_1$:

$$x_1 = x_0 - \frac{f(x_0)}{f'(x_0)}$$

$$x_1 = 2 - \frac{f(2)}{f'(2)}$$

$$x_1 = 2 - \frac{-1}{4}$$

$$x_1 = 2 + \frac{1}{4}$$

$$x_1 = \frac{8}{4} + \frac{1}{4}$$

$$x_1 = \frac{9}{4}$$

According to Newton's Method, $x_1 = \frac{9}{4} = 2.25$ is a better estimate of the positive root of $f(x)$ than the original guess of $x_0 = 2$. For an even better guess, substitute $x_1 = \frac{9}{4}$ into Newton's Method to calculate x_2:

$$x_2 = x_1 - \frac{f(x_1)}{f'(x_1)}$$

$$= \frac{9}{4} - \frac{f(9/4)}{f'(9/4)}$$

$$= \frac{9}{4} - \frac{(9/4)^2 - 5}{2(9/4)}$$

$$= \frac{9}{4} - \frac{81/16 - 5}{18/4}$$

$$= \frac{9}{4} - \frac{81/16 - 80/16}{9/2}$$

$$= \frac{9}{4} - \frac{1/16}{9/2}$$

KELLEY'S CAUTIONS

If, for some reason, the iterations of Newton's Method produce values that are getting farther apart, rather than closer together, then you started with a lousy initial seed value. Pick a value closer to the x-intercept and start over.

Time out! These fractions are getting ugly, and they only get uglier with every iteration. I think it's time to cut our losses and type this into a calculator to finish:

$$_2 = \frac{9}{4} - \frac{2}{144}$$

$$= \frac{322}{144}$$

$$\approx 2.23611$$

After two iterations of Newton's Method, you have calculated an estimated root of 2.23611 for $f(x)$. You could go on, getting closer and closer to the root, but I think we all have better things to do.

YOU'VE GOT PROBLEMS

Problem 1: Given $x_0 = 2$, apply Newton's Method to calculate x_1 and approximate the root of $g(x) = \sqrt{x+2} - 3$.

Evaluating Limits: L'Hôpital's Rule

Way, way back, many chapters ago, in a galaxy far, far away, you were stressed about limits. Since then, you've had a whole lot more to stress about, so it's high time we destressed you a bit. Little did you know that as you were plugging away, learning derivatives, you also learned a terrific shortcut for finding limits. This shortcut (called L'Hôpital's Rule) can be used to find limits that, after substitution, are in indeterminate form.

✏️ **CRITICAL POINT**

L'Hôpital's Rule can only be used to calculate limits that are indeterminate (i.e., the value cannot immediately be found). The most common indeterminate forms are $\frac{\pm\infty}{\pm\infty}$, $0/0$, and $0 \cdot \infty$.

To show just how useful L'Hôpital's Rule is, we'll return briefly to Chapter 6 and fish out two limits we couldn't previously calculate by hand. These two limits will comprise the next example. The first limit we could only memorize (but couldn't justify via any of our methods at the time). We calculated the second limit using a little trick (comparing degrees for limits at infinity), but that method was a trick only. We had no proof or justification for it at all. Finally, a little pay dirt for the curious at heart.

L'Hôpital's Rule: If $h(x) = \frac{f(x)}{g(x)}$ and $\lim\limits_{x \to c} h(x)$ is in indeterminate form (e.g., $\frac{0}{0}$ or $\frac{\infty}{\infty}$), then $\lim\limits_{x \to c} h(x) = \lim\limits_{x \to c} \frac{f'(x)}{g'(x)}$. In other words, take the derivatives of the numerator and denominator separately (not via the Quotient Rule) and substitute in c gain to find the limit.

🪜 **KELLEY'S CAUTIONS**

You can only use L'Hôpital's (pronounced *low-pee-TOWELS*) Rule if you have indeterminate form after substituting—it will not work for other, more common, limits.

Example 2: Calculate both of the following limits using L'Hôpital's Rule:

$$(a)\ \lim_{x \to 0} \frac{\sin x}{x}$$

Solution: If you substitute in $x = 0$, you get $\frac{\sin 0}{0} = \frac{0}{0}$, which is in indeterminate form. So apply L'Hôpital's Rule by taking the derivative of $\sin x$ (which is $\cos x$) and the derivative of x (which is 1) and replacing those pieces with their derivatives:

$$\lim_{x \to 0} \frac{\cos x}{1}$$

Now the substitution method won't give you $\frac{0}{0}$. In fact, substituting gives you cos 0, which equals 1. You learned that 1 was the answer in Chapter 6, but now you know why.

$$\text{(b) } \lim_{x \to \infty} \frac{5x^3 + 4x^2 - 7x + 4}{2 + x - 6x^2 + 8x^3}$$

Solution: If you plug in $x = \infty$ for all the x's you get a huge number on top divided by a huge number on the bottom, or $\frac{\infty}{\infty}$, which is in indeterminate form. Apply L'Hôpital's Rule:

$$\lim_{x \to \infty} \frac{15x^2 + 8x - 7}{1 - 12x + 24x^2}$$

Uh-oh. Substitution *still* gives you $\frac{\infty}{\infty}$. Never fear! Keep applying L'Hôpital's Rule until substituting gives you a nonindeterminate answer:

$$\lim_{x \to \infty} \frac{30x + 8}{-12 + 48x}$$
$$= \lim_{x \to \infty} \frac{30}{48}$$

Once there are no more x's in the problem, no substitution is necessary, and the answer falls out like a ripe fruit.

$$\lim_{x \to \infty} \frac{30}{48} = \frac{30}{48}$$
$$= \frac{5}{8}$$

You might remember this problem—it was Example 6 in Chapter 6. You used a different method to compute the limit then, but you got the same answer: $\frac{5}{8}$.

YOU'VE GOT PROBLEMS

Problem 2: Evaluate $\lim_{x \to \infty} \left(x^{-2} \cdot \ln x \right)$ using L'Hôpital's Rule. *Hint:* begin by writing the expression as a fraction.

More Existence Theorems

Man has struggled for centuries to define life and to determine what, exactly, defines existence. Descartes once mused, "I think; therefore, I am," suggesting that thought defined existence. Most calculus students go one step further, lamenting, "I am in mental anguish; therefore, I am in calculus." Philosophy aside, the next two theorems don't try to answer such deep questions; they simply state that something exists, and that's good enough for them.

The Mean Value Theorem

This neat little theorem gives an explicit relationship between the average rate of change of a function (i.e., the slope of a secant line) and the instantaneous rate of change of a function (i.e., the slope of a tangent line). Specifically, it guarantees that at some point on a closed interval, the tangent line will be parallel to the secant line for that interval (see Figure 13.2).

> **CRITICAL POINT**
>
> It's called the *Mean* Value Theorem because a major component of it is the average (or mean) rate of change for the function. It has no twin called the Kind Value Theorem.

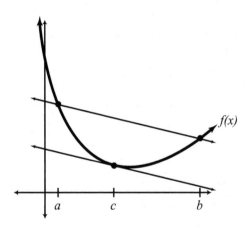

Figure 13.2
Here, the secant line is drawn connecting the endpoints of the closed interval [a,b], at x = c, which is on that interval; the tangent line is parallel to the secant line.

Mathematically, parallel lines have equal slopes. Therefore, there is always some place on an interval where a continuous function is changing at exactly the same rate it's changing on average for the entire interval. Here's the theorem in math jibber jabber:

The Mean Value Theorem: If a function $f(x)$ is continuous and differentiable on the closed interval $[a,b]$, then there exists a value c between a and b such that $f'(c) = \frac{f(b)-f(a)}{b-a}$. In other words, a value c is guaranteed to exist such that the derivative there ($f'(c)$) is equal to the slope of the secant line for the interval $[a,b]$ $\left(\frac{f(b)-f(a)}{b-a} \right)$.

The Mean Value Theorem makes good sense. Think of it like this: if, on a 2-hour car trip, you averaged 50 miles per hour, then (according to the Mean Value Theorem) at least once during the trip, your speedometer actually read 50 mph.

Example 3: At what x-value(s) on the interval $[-2,3]$ does the graph of $f(x) = x^2 + 2x - 1$ satisfy the Mean Value Theorem?

Solution: The function is continuous and differentiable because there are no domain restrictions. Somewhere, the derivative must equal the secant slope, so start by finding the derivative of $f(x)$:

$$f'(x) = 2x + 2$$

That was easy. Now find the secant slope over the interval $[-2,3]$. To calculate it, first plug -2 and 3 into the function to get the secant's endpoints, $(-2,-1)$ and $(3,14)$:

$$\frac{y_2 - y_1}{x_2 - x_1} = \frac{14 - (-1)}{3 - (-2)} = \frac{15}{5} = 3$$

Therefore, at some point on the interval, the derivative, $f'(x) = 2x + 2$, and the secant slope you calculated, 3, must be equal:

$$2x + 2 = 3$$
$$2x = 3 - 2$$
$$2x = 1$$
$$x = \frac{1}{2}$$

Look at the graph of $f(x)$ in Figure 13.3 to verify that the tangent line at $x = \frac{1}{2}$ is parallel to the secant line connecting $(-2,-1)$ and $(3,14)$.

Problem 3: Given the function $g(x) = \frac{1}{x}$, find the x-value that satisfies the Mean Value Theorem on the interval $\left[\frac{1}{4}, 1\right]$.

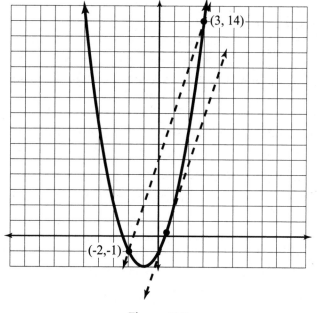

Figure 13.3
Equal secant and tangent slopes result in parallel secant and tangent lines.

Rolle's Theorem

Rolle's Theorem is a specific case of the Mean Value Theorem. It says that if the slope of a function's secant line is 0 (in other words, the secant line is horizontal because the endpoints of the interval are located at the exact same height on the graph), then somewhere on that interval, the tangent slope will also be 0. Because you already understand the Mean Value Theorem, this isn't new information. Our previous theorem guaranteed the lines would have the same slope no matter what the secant slope was. Here's how Rolle's Theorem is defined mathematically:

Rolle's Theorem: If a function $f(x)$ is continuous and differentiable on a closed interval $[a,b]$ and $f(a) = f(b)$, then there exists a c between a and b such that $f'(c) = 0$.

Let's prove this with the Mean Value Theorem—it guarantees that the secant slope will equal the tangent slope somewhere on $[a,b]$. The secant slope connecting the points $(a, f(a))$ and $(b, f(b))$ is $\frac{f(b)-f(a)}{b-a}$, but because the theorem states that $f(a) = f(b)$, this fraction becomes $\frac{0}{b-a} = 0$. Therefore, the slope of the secant line is 0. According to the Mean Value Theorem, $f'(x)$ has to equal 0 somewhere inside the interval, at a point Rolle's Theorem calls c.

Related Rates

Related rates problems are among the most popular problems (for teachers) and feared problems (for students) in calculus. You can tell if a given problem is a related rates problem because it will contain wording like "how quickly is … changing?" Basically you're asked to figure out how quickly one variable in a problem is changing if you know how quickly another variable is changing. No two problems will be alike, but the procedure is exactly the same for all problems of this type, and they actually become sort of fun once you get used to them.

Let's walk through a classic related rates problem: a ladder-sliding-down-the-side-of-a-house dilemma. The only step that will differ between this and any other related rates problem is the very first one: finding an equation that characterizes the situation. Once you get past that initial step, everything is smooth sailing.

Example 4: Goofus and Gallant (of *Highlights* magazine fame) are painting my house. Whereas Gallant properly secured his 13-foot ladder before climbing it, Goofus did not, and as he climbs his ladder, it slides down the side of the house at a constant rate of 2 feet/second. How quickly is the base of the ladder sliding horizontally away from the house when the top of the ladder is 5 feet from the ground?

Solution: You can tell this is a related rates problem because it's asking you to find how quickly something is changing or moving. I always start these by drawing a picture of the situation (see Figure 13.4).

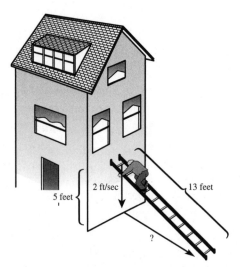

Figure 13.4
Recipe for disaster: the 13-foot ladder, with its top only 5 feet from the ground, and Goofus heroically clinging to it.

You need to pick an equation that represents the situation. Notice that the ladder, the house, and the ground make a right triangle; the problem gives you information about the lengths of the legs of a right triangle. Therefore, you should use the Pythagorean Theorem as your primary equation, as it relates the lengths of the sides of a right triangle. To make it easier to visualize, I will strip away all of the extraneous visual information:

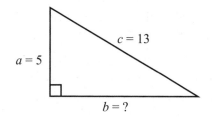

Figure 13.5
Goofus's predicament, minus the clever illustration.

KELLEY'S CAUTIONS

> Remember, you won't use the Pythagorean Theorem for every related rates problem. You'll have to pick your primary equation based on the situation. Look at Problem 3 in the "You've Got Problems" sidebar earlier in this chapter for a different example.

According to Figure 13.5 (and the Pythagorean Theorem), you know that $a^2 + b^2 = c^2$. Warning: don't plug in any values you know (like $a = 5$) until you complete the next step, which is differentiating everything with respect to t:

$$2a\frac{da}{dt} + 2b\frac{db}{dt} = 2c\frac{dc}{dt}$$

You might be wondering, "What does $\frac{da}{dt}$ mean?" It represents how quickly a is changing. The problem tells you that the ladder is falling, so side a is actually getting smaller at a rate of 2 ft/sec, so write $\frac{da}{dt} = -2$. At this moment, you have no idea what $\frac{db}{dt}$ equals, because that's the quantity you're looking for. However, you do know that $\frac{dc}{dt} = 0$, because c (the length of the ladder) will not change as it slides down the house.

KELLEY'S CAUTIONS

> If a variable is decreasing in size, its accompanying rate must be negative. In Example 4, because a is decreasing at 2 ft/sec, $\frac{da}{dt} = -2$, not 2.

Now you know most of the variables in the equation. In fact, you can even calculate $b = 12$ using the Pythagorean Theorem, knowing that the other sides of the triangle are 5 and 13. So plug in everything you know:

$$2 \cdot 5 \cdot \left(-2\right) + 2 \cdot 12 \cdot \tfrac{db}{dt} = 2 \cdot 13 \cdot 0$$

$$-20 + 24 \tfrac{db}{dt} = 0$$

All you have to do is solve for $\tfrac{db}{dt}$, and you're finished:

$$24 \tfrac{db}{dt} = 20$$

$$\tfrac{db}{dt} = \tfrac{20}{24}$$

$$\tfrac{db}{dt} = \tfrac{5}{6} \text{ ft/sec}$$

Therefore, b is increasing at a rate of $\tfrac{5}{6}$ ft/sec, and that's how quickly the base of the ladder is sliding away from the house.

> **KELLEY'S CAUTIONS**
>
> If you're wondering where all those $\tfrac{da}{dt}$'s and $\tfrac{db}{dt}$'s are coming from, flip back to Chapter 10.

Here are the steps to completing a related rates problem:

1. Construct an equation containing all the necessary variables.

2. Before substituting any values, differentiate the entire equation with respect to t.

3. Plug in values for all the variables except the one for which you're solving.

4. Solve for the unknown variable.

Example 5: If air leaks out of a spherical balloon at a rate of 2 in³/hour, how quickly is the balloon's radius decreasing (in inches/hour) when its volume is $\tfrac{4,000\pi}{3}$ in³? *Hint:* the formula for the volume of a sphere is $V = \tfrac{4}{3}\pi r^3$.

Solution: You're asked to calculate the rate the radius is decreasing. If r represents the radius, that means you're looking for $\tfrac{dr}{dt}$. You are told that the air is leaking out of the balloon, which means that the volume of the balloon is decreasing $\tfrac{dV}{dt} = -2$ in³/hour.

Take the derivative of the volume formula for a sphere, with respect to t:

$$V = \tfrac{4}{3}\pi r^3$$

$$\tfrac{dV}{dt} = \tfrac{4}{3}\pi \cdot 3 \cdot r^2 \cdot \tfrac{dr}{dt}$$

Note that you should treat π like any other number—it is part of the coefficient of the term. When you apply the Power Rule for derivatives, you multiply the coefficient $\left(\frac{4}{3}\pi\right)$ by the original exponent (3) and then subtract 1 from the exponent ($3 - 1 = 2$).

$$\frac{dV}{dt} = \frac{4}{\cancel{3}}\pi \cdot \cancel{3} \cdot r^2 \cdot \frac{dr}{dt}$$

$$\frac{dV}{dt} = 4\pi r^2 \cdot \frac{dr}{dt}$$

To solve for $\frac{dr}{dt}$ and complete the problem, you will need to know the value of r. To calculate it, return to the original formula for the volume of a sphere and calculate the radius of a sphere that has a volume of $\frac{4,000\pi}{3}$ in³:

$$V = \frac{4}{3}\pi r^3$$

$$\frac{4000\pi}{3} = \frac{4}{3}\pi r^3$$

Multiply both sides of the equation by 3 to eliminate the fractions:

$$4000\pi = 4\pi r^3$$

$$\frac{4000\pi}{4\pi} = r^3$$

$$1000 = r^3$$

$$\sqrt[3]{1000} = \sqrt[3]{r^3}$$

$$10 = r$$

Now substitute $r = 10$ and $\frac{dV}{dt} = -2$ into the derivative you calculated earlier and solve for $\frac{dr}{dt}$:

$$\frac{dV}{dt} = 4\pi r^2 \cdot \frac{dr}{dt}$$

$$-2 = 4\pi\left(10\right)^2 \cdot \frac{dr}{dt}$$

$$-2 = 400\pi \frac{dr}{dt}$$

$$\frac{-2}{400\pi} = \frac{dr}{dt}$$

$$-\frac{1}{200\pi} = \frac{dr}{dt}$$

The radius of the spherical balloon is decreasing at a rate of $\frac{1}{200\pi} \approx 0.001592$ inches/hour.

Optimization

Even though optimization is (arguably) the most feared of all differentiation applications, I have never understood why. When you're looking for the biggest or smallest something can get (i.e., optimizing), all you have to do is create a formula representing that quantity and then find the relative extrema using wiggle graphs. You've been doing these things for a while now, so don't get freaked out unnecessarily. To explore optimization, we'll again examine a classic calculus problem that has haunted students like you for years and years.

Example 6: If you create a box by cutting congruent squares from the corners of a piece of paper measuring 11 by 14 inches, give the dimensions of the box with the largest possible volume. (Assume that the box has no lid.)

Solution: Back in Chapter 1, I hinted about how to create a box out of a flat piece of paper. Try it for yourself. Place a rectangular sheet of paper in front of you and cut congruent squares from the corners. You'll end up with smaller rectangles along the sides of your paper. Fold these up, toward you, along the seam created by the inner sides of the recently removed squares. Can you see how the remaining rectangles correspond to the dimensions of the box (see Figure 13.6)?

I have labeled the sides of the corner squares as x in Figure 13.6. Once you cut out those squares, the length of the top and bottom is $11 - 2x$, because it was 11 inches and you removed two lengths, each measuring x inches. Similarly, the sides of the box will measure $14 - 2x$ inches.

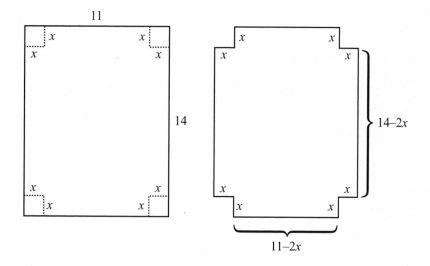

Figure 13.6
The height of the box will be x inches, because the side length of the cut-out squares dictates how deeply to fold the paper.

Now that you have a good idea what is happening visually, let's get hip-deep in the math. You are trying to make the largest possible volume, so your primary equation should be for the volume for this box. The volume for any box like this is $V = l \cdot w \cdot h$, where l = length, w = width, and h = height. Plug in the correct values for l, w, and h:

$$V = l \cdot w \cdot h$$
$$V = \left(14 - 2x\right)\left(11 - 2x\right)x$$
$$V = \left(154 - 50x + 4x^2\right)x$$
$$V = 4x^3 - 50x^2 + 154x$$

KELLEY'S CAUTIONS

In Example 7, consider only values of x between 0 and 5.5. Why? Well, if x is less than 0, you're not cutting out any squares, and if x is greater than 5.5, then the $(11 - 2x)$ width of your box becomes 0 or smaller, and that's just not allowed. A real-life box must have *some* width.

If you plug in any x, this function gives you the volume of the box generated when squares of side x are cut out. Cool, eh? You want to find the value of x that makes V the largest, so find the value guaranteed by the Extreme Value Theorem. Take the derivative with respect to x and do a wiggle graph (see Figure 13.7), just like you did in Chapter 11.

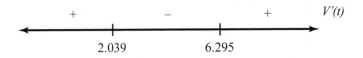

Figure 13.7
The wiggle graph of V'. *A relative maximum occurs at* x = *2.039.*

$$V' = 12x^2 - 100x + 154$$

Set the derivative equal to 0, and while you're at it, divide everything by 2 in order to simplify the coefficients a bit.

$$6x^2 - 50x + 77 = 0$$

You can use your calculator or the quadratic formula to solve the equation, and you get solutions $x \approx 2.039$ and $x \approx 6.295$. Although $x = 6.295$ appears to be a minimum, because the function changes from decreasing to increasing there, the answer doesn't make sense. See the "Kelley's Cautions" sidebar for more details.

KELLEY'S CAUTIONS

As you plug the variables into the primary equation, your goal should be to have only one main variable. In Example 6, you change *l, w,* and *h* so they all contain only one variable (*x*). Don't worry that *V* is a variable—you don't deal with the left side of the equation at all.

The maximum volume is reached when $x = 2.039$ (because V' changes from positive to negative there, meaning that V goes from increasing to decreasing), so the optimal dimensions are 2.039 inches by 6.922 inches by 9.922 inches (x, $11 - 2x$, and $14 - 2x$, respectively).

Here are the steps for optimizing functions:

1. Construct an equation in one variable that represents what you are trying to maximize.

2. Find the derivative with respect to the variable in the problem and draw a wiggle graph.

3. Verify your solutions as the correct extrema type (either maximum or minimum) by viewing the sign changes around it in the wiggle graph.

Example 7: A company wishes to package its fruit in a cylindrical aluminum can with a volume of 30 in³. Identify the radius and height of the can that minimizes the aluminum needed, reporting each accurate to the thousandths place.

Solution: The aluminum is used to fabricate the cylinder of the can. If you are trying to minimize the amount of aluminum used, you are actually trying to minimize the surface area of the can. Less surface area means less aluminum.

Consider Figure 13.8, which illustrates how to calculate the surface area of a cylinder. The total surface area is equal to the surface area of the circular top and bottom plus the surface area of the sides of the can.

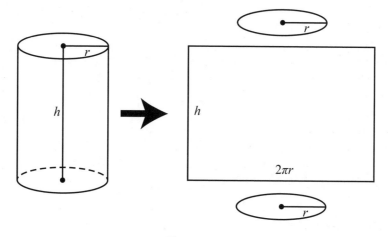

Figure 13.8

A cylindrical can has three parts. Two circles with radius r *form the top and bottom. A rectangle with length 2πr and height* h *forms the sides of the can.*

The sides of the can are formed by a single rectangle wrapped in a circle. Therefore, the length of the rectangle is equal to the circumference of the circle ($2\pi r$), and the height of the rectangle matches the height of the can. Now you can write the formula for S, the surface area of the can:

$$S = \text{area of top and bottom} \ + \text{area of sides}$$
$$S = \pi r^2 + \pi r^2 \qquad\qquad +2\pi rh$$
$$S = 2\pi r^2 + 2\pi rh$$

Unfortunately, there are too many variables present. You need a function that expresses surface area in terms of r or h, but not r and h. Luckily, there is more information in the problem, including something you haven't used yet. The volume of the can must be 30 in³. Set the formula for the volume of a cylinder equal to 30 and solve for one of the variables. It is easier to solve for h than to solve for r:

$$\text{Volume of cylinder} = 30$$
$$\pi r^2 h = 30$$
$$h = \frac{30}{\pi r^2}$$

If you substitute this value for h into the surface area function, you get a function completely devoid of h's, containing only r's:

$$S = 2\pi r^2 + 2\pi rh$$

$$S = 2\pi r^2 + 2\pi r \left(\frac{30}{\pi r^2} \right)$$

$$S = 2\pi r^2 + \frac{60\pi r}{\pi r^2}$$

$$S = 2\pi r^2 + \frac{60}{r}$$

Take the derivative with respect to r. It may help to rewrite $\frac{60}{r}$ as $60r^{-1}$ so you can use the Power Rule:

$$S = 2\pi r^2 + 60r^{-1}$$

$$S' = 4\pi r - 60r^{-2}$$

$$S' = 4\pi r - \frac{60}{r^2}$$

Set $S' = 0$ and solve for r by cross-multiplying:

$$0 = 4\pi r - \frac{60}{r^2}$$

$$\frac{60}{r^2} = \frac{4\pi r}{1}$$

$$4\pi r^3 = 60$$

$$r^3 = \frac{60}{4\pi}$$

$$r^3 = \frac{15}{\pi}$$

$$r = \sqrt[3]{\frac{15}{\pi}}$$

$$r \approx 1.683890301 \text{ inches}$$

You can verify that this value of r is a minimum using a wiggle graph if you wish. One more task: it's time to calculate the corresponding height. Substitute this value of r into the formula you solved for height only moments ago:

$$h = \frac{30}{\pi r^2}$$

$$h = \frac{30}{\pi \left(1.683890301 \right)^2}$$

$$h \approx 3.367780602$$

The cylinder with a volume of 30 in^3 that uses a minimum amount of materials for construction has a radius of 1.684 inches and a height of 3.368 inches.

YOU'VE GOT PROBLEMS

Problem 5: What is the minimum product you can achieve from two real numbers, if one of them is three less than twice the other?

The Least You Need to Know

- Newton's Method is an iterative formula used to approximate the roots of a function based on the values of the function and its derivative.

- L'Hôpital's Rule is a shortcut to finding limits that are indeterminate when you try to solve them using substitution.

- The Mean Value Theorem guarantees that the secant slope on an interval will equal the tangent slope somewhere on that interval—i.e., the average rate of change must somewhere be equal to the instantaneous rate of change.

- You can determine how quickly a variable is changing in an equation if you know how quickly the other variables in the equation are changing.

- The first derivative can help you determine where a function reaches its optimal values.

The Integral

For those of you with a good background in superhero (or *Seinfeld*) lore, you'll know what I mean by Bizarro world. In Bizarro world, everything is the opposite of this world; good means bad, up means down, and right means left. Because Superman is smart in our world, Bizarro Superman is stupid. Well, integrals are Bizarro derivatives. Deriving takes us from a function to an expression describing its rate of change, but integrating takes us in the opposite direction—from the rate of change back to the original function.

Even though integrating is simply the opposite of deriving, you might think that its usefulness would be limited. You'd be wrong. There are just about as many applications for integrals as there are for derivatives, but they are completely different in nature. Instead of finding rates of change, we'll calculate area, volume, and distance traveled. We'll also explore the Fundamental Theorem of Calculus, which explains the exact relationship between integrals and the area beneath a curve. It's surprising how straightforward that relationship is and how dang useful it can be.

CHAPTER 14

Approximating Area

Have you ever seen the movie *Speed* with Keanu Reeves and Sandra Bullock? If not, here's a recap. Everyone's trapped in this city bus, which will explode if the speedometer goes below 50 mph. So, you've got this killer, runaway bus that's flying around the city and can't stop—the perfect breeding ground for destruction, disaster, high drama, mayhem, and a budding romance between the movie's two stars. (Darn that Keanu. Talk about being in the right place at the right time ….)

By now, you probably feel like you're on that bus. Calculus is tearing all over the place, never slowing down, never stopping, and (unfortunately) never inhabited by such attractive movie stars. The more you learn about derivatives, the more you have to remember about the things that preceded them. Just when you understand something, another (seemingly unrelated) topic pops up to confound your understanding. When will this bus slow down? Actually, the bus slows down right now.

You may feel a slight lurching in the pit of your stomach as we slow to a complete stop, and start discussing something completely and utterly different for a while. Until now, we've spent a ton of time talking about rates of change and tangent slopes. That's pretty much over. Instead, we're going to start talking about finding the area under curves. I know that's a big change, but it'll all come together in the end. For now, take a deep breath, and enjoy a much slower pace for a few chapters as we talk about something different. And if you see Sandra Bullock, tell her I said hi.

Riemann Sums

Let me begin by saying something deeply philosophical. Curves are really, really curvy. It is this inherent curviness that makes it hard to find the area beneath them. For example, take a look at the graph of $y = x^2 + 1$ in Figure 14.1 (only the interval [0,3] is pictured).

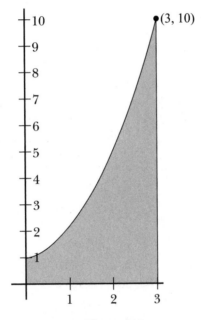

Figure 14.1

If only the shaded region between y *=* x² *+* 1 *and the* x-*axis were a square or rectangle—that would make finding the area so much easier.*

We want to try and figure out exactly how much area is represented by the shaded space. We don't have any formulas from geometry to help us find the area of such a curved figure, so we're going to need to come up with some new techniques. To start with, we're going to approximate that area using figures for which we already have area formulas. Even though it seems kind of lame, we're going to approximate the shaded area using rectangles. The process of using rectangles to approximate area is called *Riemann sums*.

 **DEFINITION**

A **Riemann sum** is an approximation of an area calculated using rectangles.

We will be exploring simple Riemann sums. Some calculus courses will explore very complicated sums, which involve crazy formulas containing sigma signs (Σ). In my opinion, these won't help you understand the underlying calculus concepts, so I omit them.

> ✏️ **CRITICAL POINT**
>
> When I say we are looking for the area *beneath* the curve, I actually mean the area *between* the curve and the x-axis; otherwise, the area beneath a curve would almost always be infinite. You can always assume that you are finding the area between the curve and the x-axis unless the problem states otherwise.

Right and Left Sums

I'm going to approximate that shaded area beneath $y = x^2 + 1$ using three rectangles. Because I'm only finding the area on the x-interval $[0,3]$, that means I'll be using three rectangles, each of width 1. (If I had been using six rectangles on an interval of length 3, each rectangle would have width $\frac{1}{2}$.) How high should I make each rectangle? Well, I choose to use a *right sum*, which means that the rectangles will have the height reached by the function at the right side of each interval, as pictured in Figure 14.2.

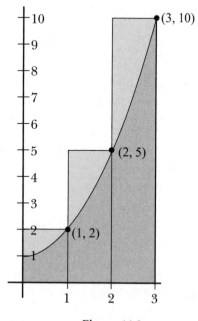

Figure 14.2

I'm using three rectangles to approximate the area on [0,1], [1,2], and [2,3].

The rectangle on [0,1] will have the height reached at the far right side of the interval (i.e., $x = 1$), which is 2. Similarly, the second rectangle is 5 units tall, because that is the height of the function at $x = 2$, the right side of its interval. Therefore, the heights of the rectangles are 2, 5, and 10, from left to right. The width of each rectangle is 1.

We can approximate the area beneath the curve by adding the areas of the three rectangles together. Because the area of a rectangle is equal to its length times its width, the total area captured by the rectangles is $1 \cdot 2 + 1 \cdot 5 + 1 \cdot 10 = 17$. Therefore, the right Riemann sum approximation with $n = 3$ rectangles is 17.

> ✏️ **CRITICAL POINT**
>
> It was easy to see that the width of every rectangle in our right sum was 1. If the width of the rectangles is not so obvious, use the width formula $x = \frac{b-a}{n}$ to calculate the width. In this formula, the interval $[a,b]$ is split up into n different rectangles, and each will have width Δx. In our right sum example, $x = \frac{3-0}{3} = 1$, because we are splitting up the interval $[0,3]$ into $n = 3$ rectangles.

Clearly, the area covered by the rectangles is much more than is beneath the curve. In fact, it looks like a lot more. This should tell you that we have got to come up with better methods later (and indeed we will). For now, let's have a go at the same area problem, but this time use four rectangles and *left sums*.

> 📖 **DEFINITION**
>
> The kind of sum you're calculating depends on how high you make the rectangles. If you use the height at each rectangle's left boundary, you're finding **left sums.** If you use the height at the right boundary of each rectangle, the result is right sums. Obviously, midpoint sums use the height reached by the function in the middle of each interval.

Example 1: Approximate the area beneath the curve $f(x) = x^2 + 1$ on the interval $[0,3]$ using a left Riemann sum with four rectangles.

Solution: To find how wide each of the four rectangles will be, use the formula $x = \frac{b-a}{n}$:

$$x = \frac{3-0}{4} = \frac{3}{4}$$

If each of the four intervals is $\frac{3}{4}$ wide, and the rectangles start at 0, then the rectangles will be defined by the intervals $\left[0, \frac{3}{4}\right], \left[\frac{3}{4}, \frac{3}{2}\right], \left[\frac{3}{2}, \frac{9}{4}\right],$ and $\left[\frac{9}{4}, 3\right]$. (This is because $0 + \frac{3}{4} = \frac{3}{4}$, $\frac{3}{4} + \frac{3}{4} = \frac{6}{4} = \frac{3}{2}$, etc.) You will be using the heights reached by the function at the left boundary of each interval. Therefore, the heights will be $f\left(0\right), f\left(\tfrac{3}{4}\right), f\left(\tfrac{3}{2}\right)$, and $f\left(\tfrac{9}{4}\right)$, as illustrated in Figure 14.3.

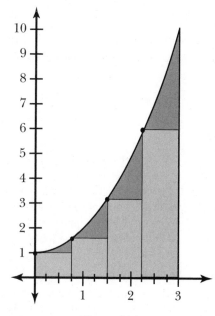

Figure 14.3

Each of the four rectangles is $\tfrac{3}{4}$ wide, and they are as high as the function f(x) = x² + 1 at the left edge of each rectangle, hence left sums is the result.

The area of each rectangle is its width times its height, so the total area is:

$$\left[\tfrac{3}{4}\cdot f\left(0\right)\right]+\left[\tfrac{3}{4}\cdot f\left(\tfrac{3}{4}\right)\right]+\left[\tfrac{3}{4}\cdot f\left(\tfrac{3}{2}\right)\right]+\left[\tfrac{3}{4}\cdot f\left(\tfrac{9}{4}\right)\right]$$

$$=\tfrac{3}{4}\left(1\right)+\tfrac{3}{4}\left(\tfrac{25}{16}\right)+\tfrac{3}{4}\left(\tfrac{13}{4}\right)+\tfrac{3}{4}\left(\tfrac{97}{16}\right)$$

$$\approx 8.906$$

This number underestimates the actual area beneath the curve, because there are large pieces of that area missed by our rectangles.

Midpoint Sums

Calculating midpoint sums is similar to calculating right and left sums. The only difference is (you guessed it) how you define the heights of the rectangles. In our ongoing example of $f(x) = x^2 + 1$ on the *x*-interval [0,3], let's say we wanted to calculate midpoint sums using (to make it easy) $n = 3$ rectangles. As before, the intervals defining the rectangles' boundaries will be [0,1], [1,2], and [2,3], and each rectangle will have a width of 1. What about the heights?

Look at the interval [0,1]. If we were using left sums, the height of the rectangle would be $f(0)$. Using right sums, it'd be $f(1)$. However, we're using midpoint sums, so you use the function value at the midpoint of the interval, which in this case is $\frac{1}{2}$. Therefore, the height of the rectangle is $f\left(\frac{1}{2}\right)$. If you apply this to all three intervals, the midpoint Riemann approximation of the area would be:

$$1 \cdot f\left(\frac{1}{2}\right) + 1 \cdot f\left(\frac{3}{2}\right) + 1 \cdot f\left(\frac{5}{2}\right) = \frac{47}{4} = 11.75$$

YOU'VE GOT PROBLEMS

Problem 1: Approximate the area beneath the curve $g(x) = -\cos x$ on the interval $\frac{\pi}{2}, \frac{3\pi}{2}$ using $n = 4$ rectangles and (1) left sums, (2) right sums, and (3) midpoint sums.

The Trapezoidal Rule

Unfortunately, unless you use a ton of rectangles, Riemann sums are just not all that accurate. The Trapezoidal Rule, however, is often a more accurate way to approximate area beneath a curve. Instead of constructing rectangles, this method uses small trapezoids. In effect, these trapezoids look the same as their predecessor rectangles near their bases, but completely different at the top. To construct the trapezoids, you mark the height of the function at the beginning and end of the width interval (which is still calculated by the formula $x = \frac{b-a}{n}$) and connect those two points. Figure 14.4 shows how the Trapezoidal Rule approximates the area beneath our favorite function in the whole world, $y = x^2 + 1$.

CRITICAL POINT

This is going to freak you out. Remember how the left and right sums offset one another when we approximated the area beneath $y = x^2 + 1$—one too big and the other too small? Well, the Trapezoidal Rule (with n trapezoids) is exactly the average of the left and right sums (with n rectangles). We already know that the right sum of $y = x^2 + 1$ (with $n = 3$) is 17. You can find the corresponding left sum to be 8. If you calculate the Trapezoidal Rule approximation (with $n = 3$ trapezoids), you get 12.5, which is the average of 8 and 17.

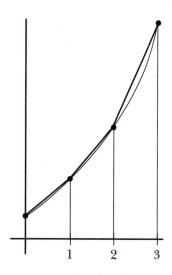

Figure 14.4
The "tops" of our approximating shapes are no longer parallel to the x-axis. Instead they connect the
function's heights at the interval endpoints.

There's a lot less room for error with this rule, and it's actually just as easy to use as Riemann sums were. One difference—this one requires that you memorize a formula.

CRITICAL POINT

> If you're dying to know the actual area beneath $y = x^2 + 1$ on the interval [0,3], it is exactly 12. Of our approximations so far, the midpoint sum came the closest (even though we used only three rectangles with this method but four with left sums).

The Trapezoidal Rule: The approximate area beneath a continuous curve $f(x)$ on the interval $[a,b]$ using n trapezoids equals:

$$\frac{b-a}{2n}\left[f(a)+2f(x_1)+2f(x_2)+2f(x_3)+\cdots+2f(x_{n-1})+f(b)\right]$$

In practice, you pop the correct numbers into the fraction at the beginning and then evaluate the function at every interval boundary. Except for the endpoints, you'll multiply all the values by 2.

The area of any trapezoid is one-half the height times the sum of the bases (the bases are the parallel sides). For the trapezoid in Figure 14.5, the area is $\frac{1}{2}h(b_1 + b_2)$. You may not be used to seeing trapezoids tipped on their side like this—in geometry, the bases are usually horizontal, not vertical. The reason you see all those 2's in the Trapezoidal Rule is that every base is used twice for consecutive trapezoids except for the bases at the endpoints.

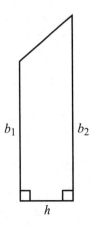

Figure 14.5
Our approximate trapezoids are simply right trapezoids shoved onto their sides, with bases b_1 *and* b_2,
and height h.

CRITICAL POINT

There's another way to get better approximations using Riemann sums. If you increase the number of rectangles you use, the amount of error decreases. However, the amount of calculating you have to do increases. Eventually, we'll find a way to obtain the *exact* area without much work at all. The way is rooted in Riemann sums, but uses an *infinite* number of rectangles in order to eliminate any error completely.

Let's go straight into an example, and you'll see that the Trapezoidal Rule is not very hard at all. Just for grins, let's use $f(x) = x^2 + 1$ yet again to see if the Trapezoidal Rule can beat out our current best estimate of 11.75 given by the midpoint sum.

Example 2: Approximate the area beneath $f(x) = x^2 + 1$ on the interval $[0,3]$ using the Trapezoidal Rule with $n = 5$ trapezoids.

Solution: Because you are using five trapezoids, you need to determine how wide each will be, so apply the Δx formula:

$$x = \frac{b-a}{n} = \frac{3-0}{5} = \frac{3}{5}$$

KELLEY'S CAUTIONS

Although the Trapezoidal Rule's formula contains the expression $\frac{b-a}{2n}$, you still use the formula $\frac{b-a}{n}$ to find the width of the trapezoids. Don't get them confused—they are separate formulas.

Therefore, the boundaries of the intervals will start at $x = 0$ and progress in steps of $\frac{3}{5}$: $0, \frac{3}{5}, \frac{6}{5}, \frac{9}{5}, \frac{12}{5}$, and 3. These numbers belong in the formula as a, x_1, x_2, x_3, x_4, and b. So according to the Trapezoidal Rule, the area is approximately:

$$\frac{3-0}{2(5)}\left[f(0)+2f\left(\tfrac{3}{5}\right)+2f\left(\tfrac{6}{5}\right)+2f\left(\tfrac{9}{5}\right)+2f\left(\tfrac{12}{5}\right)+f(3)\right]$$

$$=\frac{3}{10}\left[1+2\cdot\tfrac{34}{25}+2\cdot\tfrac{61}{25}+2\cdot\tfrac{106}{25}+2\cdot\tfrac{169}{25}+10\right]$$

$$=\frac{609}{50}$$

$$=12.18$$

This is actually the closest approximation yet, although it is a bit too big. Had this curve been concave down instead of up, the result would have underestimated the area. Can you see why? The teeny bit of error would have been outside, rather than inside, the curve.

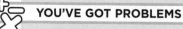

YOU'VE GOT PROBLEMS

Problem 2: Approximate the area beneath $y = \sin x$ on the interval $[0,\pi]$ using the Trapezoidal Rule with $n = 4$ trapezoids.

Simpson's Rule

Our final area-approximating tool is Simpson's Rule. Geometrically, it creates tiny little parabolas (rather than the slanted trapezoidal interval roofs) to wrap even closer around the function we're approximating. The formula is astonishingly similar to the Trapezoidal Rule, but here's the catch: you can only use an even number of subintervals.

Simpson's Rule: The approximate area under the continuous curve $f(x)$ on the closed interval $[a,b]$ using an even number of subintervals, n, is:

$$\frac{b-a}{3n}\left[f(a)+4f(x_1)+2f(x_2)+\cdots+2f(x_{n-2})+4f(x_{n-1})+f(b)\right]$$

In this formula, the outermost terms get multiplied by nothing. However, beginning with the second term, you multiply consecutive terms by 4, then 2, then 4, then 2, etc. Make sure you always start with 4, though. Back to Old Faithful, $f(x) = x^2 + 1$, for an example.

Example 3: Approximate the area beneath that confounded function $f(x) = x^2 + 1$ on the closed interval $[0,3]$, this time using Simpson's Rule and $n = 6$ subintervals.

Solution: Some quick calculating tells us that our subintervals will have the width of $x = \frac{b-a}{n} = \frac{3-0}{6} = \frac{1}{2}$. Now, to the formula we go:

$$\frac{b-a}{3n}\left[f(0)+4f\left(\tfrac{1}{2}\right)+2f(1)+4f\left(\tfrac{3}{2}\right)+2f(2)+4f\left(\tfrac{5}{2}\right)+f(3)\right]$$

Remember to multiply $f\left(\frac{1}{2}\right)$ by 4, the next term by 2, etc. However, the first and last terms get no additional coefficient:

$$\frac{3-0}{3 \cdot 6}\left[1+4\cdot\frac{5}{4}+2\cdot2+4\cdot\frac{13}{4}+2\cdot5+4\cdot\frac{29}{4}+10\right]$$
$$=\frac{1}{6}\left[1+5+4+13+10+29+10\right]$$
$$=\frac{1}{6}\left(72\right)$$
$$=12$$

Whoa! Because Simpson's Rule uses quadratic approximations, and this is a quadratic function, you get the exact answer. This only happens for areas beneath quadratic equations, though.

YOU'VE GOT PROBLEMS

Problem 3: Approximate the area beneath $y=\frac{1}{x}$ on the interval [1,5] using Simpson's Rule with $n=4$ subintervals.

The Least You Need to Know

- Riemann sums use rectangles to approximate the area beneath a curve; the heights of these rectangles are based on the height of the function at the left end, right end, or midpoint of each subinterval.

- The width of each subinterval in all the approximating techniques is $x=\frac{b-a}{n}$.

- The Trapezoidal Rule is the average of the left and right sums, and usually gives a better approximation than either does individually.

- Simpson's Rule uses intervals topped with parabolas to approximate area; therefore, it gives the exact area beneath quadratic functions.

CHAPTER 15

Antiderivatives

Are you a little perplexed? Probably. We spent the first 13 chapters of the book discussing complex mathematical procedures, and then suddenly and without warning we're calculating the area of rectangles in Chapter 14. Kind of a letdown, I know. Most people have this terrifying view of calculus, and assume that everything in it is impossible to understand; they are usually surprised to be calculating simple areas this deep in the course.

In this chapter, we'll find *exact* areas beneath a curve. We'll also uncover one of the most fascinating mathematical relationships of all time: the area beneath a curve is related to the curve's antiderivative. You heard me right—antiderivative. After all this time learning how to find the derivative of a function, now we're going to go backward and find the antiderivative. Before, we took $f(x) = x^3 - 2x^2$ and got $f'(x) = 3x^2 - 4x$; now, we're going to start with the derivative and figure out the original function.

It's a whole new ballgame, and we're going to learn everything from the first half of the course in reverse. For those of us who always seem to do things backwards, this should come as a welcome change! Sound exciting? Sound painful? It's a little from column A and a little from column B.

In This Chapter

* "Un-deriving" expressions
* The Power Rule for Integration
* Integrating trigonometric functions
* Don't hate, separate!
* The Fundamental Theorem: the connection to area
* The key to *u*-substitution

The Power Rule for Integration

Before we get started, let's talk briefly about what reverse differentiating means. The process of going from the expression $f'(x)$ back to $f(x)$ is called *antidifferentiation* or *integration*—both words mean the same thing. The result of the process is called an *antiderivative* or an *integral*. Basically, an antiderivative (or integral) is the opposite of a derivative.

> **DEFINITION**
>
> The **antiderivative** is the opposite of a derivative, but you probably guessed that. The derivative of x^2 is $2x$, so one of the antiderivatives of $2x$ is x^2.

Integration is denoted using a long, stretched-out letter S, like this:

$$\int 2x\,dx = x^2 + C$$

This is read "The integral of $2x$, with respect to x, is equal to x^2 plus some unknown constant" (called the *constant of integration*). This integral expression is called an *indefinite integral* because there are no boundaries on it.

Why do you have to use a constant of integration? Lots of functions have the same derivative—for example, both $h(x) = x^3 + 6$ and $j(x) = x^3 - 12$ have the same derivative, $3x^2$. Therefore, when we integrate $\int 3x^2\,dx$, you say the antiderivative is $x^3 + C$, because you have no way of knowing what constant was in the original function.

Whereas an indefinite integral has no boundaries next to the integration sign, a definite integral does. For example, $\int_1^3 2x\,dx$ is a definite integral because it contains the limits of integration 1 and 3. The result of an indefinite integral is a new expression, but the result of a definite integral is a real number. For example, $\int 2x\,dx = x^2 + C$, but $\int_1^3 2x\,dx = 8$. (You'll learn how to solve these kinds of problems soon.)

Both definite and indefinite integrals contain a "*dx*"; don't worry about this little piece—you don't have to do anything with it. Just make sure its variable matches the variable in the function (in this case, x).

That's a lot of vocabulary for now. Before you get overwhelmed, let's get into the meat of the mathematics. Remember finding simple derivatives with the Power Rule? There's a way to find simple integrals using the Power Rule for Integration. Instead of multiplying the original coefficient by the exponent and then subtracting 1 from the power, you'll *add* 1 to the power and *divide* by the new power.

CRITICAL POINT

According to the Power Rule for Integration, the integral of a constant is a linear term: $\int 8\,dx = 8x + C$. Just glue a variable onto the number and you're done.

The Power Rule for Integration: The integral of a single variable to some power is found by adding 1 to the existing exponent and dividing the variable by the new exponent:

$$\int x^n\,dx = \frac{x^{n+1}}{n+1} + C$$

Remember, you can only use the Power Rule for Integration if you are integrating a single variable to a power, just like the regular Power Rule. However, if the only thing standing in your way is a coefficient, you are allowed to yank it out of the integral to get it out of your way, as indicated in the first example.

CRITICAL POINT

You pull the coefficients out of the integrals to make the integration itself easier. As soon as the integration sign is gone, you end up multiplying that coefficient by the integral anyway, so it's not as though it "goes away" somewhere. It just hangs around, waiting for the integration to be done.

Example 1: Evaluate $\int \left(7x^3 + 6x^5 \right) dx$.

Solution: Even though there are two terms here, each is simply a variable to some power with a coefficient attached. You can actually separate addition or subtraction problems into separate integrals as follows:

$$\int 7x^3\,dx + \int 6x^5\,dx$$

Don't worry about the $\int$ or dx in the problem. They're the "bookends" of an integral expression, marking where it begins and ends; just integrate whatever falls between them. Before you can apply the Power Rule for Integration, you should "pull out" the coefficients:

$$7\int x^3\,dx + 6\int x^5\,dx$$

Now the expression in each integral looks like the one in the Power Rule for Integration theorem. Add 1 to each power and divide each variable by its new power. The integral sign and the "*dx*" will disappear, but don't forget to add "+ *C*" to the end of the problem, because all indefinite integrals require it:

$$7 \cdot \frac{x^4}{4} + 6 \cdot \frac{x^6}{6} + C$$
$$= \frac{7}{4}x^4 + x^6 + C$$

YOU'VE GOT PROBLEMS

Problem 1: Evaluate $\int \left(2x^4 + \frac{x^3}{3} + \sqrt{x} \right) dx$.

Integrating Trigonometric Functions

As with learning trigonometric derivatives, learning trigonometric integrals just means memorizing the correct formulas. If you forget them, you can actually create some of them from scratch easily (like the integral of the tangent function, as you'll see later in the chapter). However, not all of them are quite so easy to build by yourself, so I see some quality memorizing time in your not-too-distant future.

I can tell by that unhappy look on your face that the thought of more memorizing doesn't excite you. (You're going to be even unhappier if you haven't flipped ahead to the actual formulas yet— they are crazy looking.) Think back. You had to memorize the multiplication tables in elementary school, remember? This is just sort of the grandfather of the multiplication tables, but important all the same.

And now, with no further ado, here are the trigonometric functions with their antiderivatives:

- $\int \sin x \, dx = -\cos x + C$
- $\int \cos x \, dx = \sin x + C$
- $\int \tan x \, dx = -\ln|\cos x| + C$
- $\int \cot x \, dx = \ln|\sin x| + C$
- $\int \sec x \, dx = \ln|\sec x + \tan x| + C$
- $\int \csc x \, dx = -\ln|\csc x + \cot x| + C$

CRITICAL POINT

All of the integrals on the list containing a "co-" function are negative.

There are a lot of natural log functions in the list of trig integrals. That is due, in no small part, to the fact that $\int \frac{1}{x}\,dx = \ln|x| + C$, another important formula to memorize.

Here's another, while we're at it: the integral of e^x is itself, just like it was its own derivative; therefore, $\int e^x\,dx = e^x + C$. Integrating logarithmic functions is very, very tricky, so we don't even attempt that in Calculus I. We'll save that for Calculus II—that way you have something to look forward to! (Or dread. Take your pick.)

Example 2: Integrate: $\int\left(\sqrt[5]{x} + \sec x\right)dx$.

Solution: Rewrite this sum as a sum of two separate integrals:

$$\int\left(\sqrt[5]{x} + \sec x\right)dx = \int \sqrt[5]{x}\,dx + \int \sec x\,dx$$

The radical can be expressed as a fractional exponent $\left(\sqrt[5]{x^1} = x^{1/5}\right)$, which means you can apply the Power Rule for Integration:

$$= \int x^{1/5}\,dx + \int \sec x\,dx$$

$$= \frac{x^{(1/5)+1}}{(1/5)+1} + C + \int \sec x\,dx$$

$$= \frac{x^{(1/5)+(5/5)}}{(1/5)+(5/5)} + C + \int \sec x\,dx$$

$$= \frac{x^{6/5}}{6/5} + C + \int \sec x\,dx$$

$$= \frac{5}{6}x^{6/5} + C + \int \sec x\,dx$$

To complete the problem, recall that $\int \sec x\,dx = \ln|\sec x + \tan x| + C$.

$$= \frac{5}{6}x^{6/5} + \ln|\sec x + \tan x| + C$$

While technically both integrals produce a coefficient of integration (C), you can add those constants to a new (and still completely unknown) constant C. Is it weird to add C to C and get C? Yes, but remember that we have no idea what C equals, and we have no way of ever finding out in the context of indefinite integrals. For all intents and purposes, just remember to staple a single "+ C" to the end of any indefinite integral, no matter how many individual integrals you introduce along the way.

Separation

Breaking up is hard to do, but under specific circumstances, it is really quite worthwhile. Sometimes things just don't work out, and fractions have to go their separate ways. After a long, sunny time in the numerator together, terms just want a little more "me" time and some personal space. However, after all the time they've spent together, they've saved up a little bundle in the denominator, and both want to walk away with it.

The good news is, in the math world, both pieces of the numerator get a full share of the denominator—no lawyers, no haggling over how it should be broken up. Both terms of the numerator walk away with a full denominator, and are a little wiser for having gotten involved in the first place.

Back in grade school, you learned that two fractions couldn't be added unless they had the same denominator. With this knowledge, you proudly calculated things like $\frac{1}{3} + \frac{7}{3} = \frac{7+1}{3} = \frac{8}{3}$ and never looked back. Well, look at it backward for just a moment. If you are given the fraction $\frac{a+b}{c}$, you can rewrite it as $\frac{a}{c} + \frac{b}{c}$, just as you know that $\frac{8}{3} = \frac{7}{3} + \frac{1}{3}$.

Top-heavy integrals (which have lots of terms in the numerator but only one in the denominator) and other fractional integrals are occasionally easier to solve if you split the larger integral into smaller, more manageable ones. Although the original problem couldn't be solved via *u*-substitution or the Power Rule, the smaller integrals usually can.

KELLEY'S CAUTIONS

Never split the denominator of a fraction—only split the numerator. Although $\frac{1+3}{2} = \frac{1}{2} + \frac{3}{2}$, watch what happens if you flip the fraction over: $\frac{2}{1+3} \neq \frac{2}{1} + \frac{2}{3}$.

Example 3: Find $\int \frac{x^4 - 2x^3 + 5x^2 - 3x + 1}{x^2} \, dx$ using the separation technique.

Solution: This is a fraction, so the Power Rule for Integration doesn't apply, and setting the numerator or denominator equal to *u* is not going to do a whole lot for you, so *u*-substitution is out. If, however, you separate the five terms of the large numerator into five separate fractions, watch what happens:

$$\int \frac{x^4}{x^2} \, dx - 2\int \frac{x^3}{x^2} \, dx + 5\int \frac{x^2}{x^2} \, dx - 3\int \frac{x}{x^2} \, dx + \int \frac{1}{x^2} \, dx$$

When you simplify each of these fractions, you get simple integrals, each of which can be integrated via the Power Rule for Integration:

$$\int x^2 \, dx - 2\int x \, dx + 5\int dx - 3\int \frac{1}{x} \, dx + \int x^{-2} \, dx$$

$$= \frac{x^3}{3} - 2 \cdot \frac{x^2}{2} + 5 \cdot \frac{x^1}{1} - 3\ln|x| + \frac{x^{-1}}{-1} + C$$

$$= \frac{x^3}{3} - x^2 + 5x - 3\ln|x| - \frac{1}{x} + C$$

YOU'VE GOT PROBLEMS

Problem 2: Find $\int \frac{\sin x + \cos x}{\cos x} \, dx$ using the separation technique.

The Fundamental Theorem of Calculus

Finally, it's time to solve two mysteries of recent origin: how do you find exact areas under curves, and why are we even mentioning areas—isn't this chapter about integrals? It turns out that the exact area beneath a curve can be computed using a definite integral. This is one of two major conclusions, which together make up the Fundamental Theorem.

Part One: Areas and Integrals Are Related

After all the time we spent approximating it in Chapter 14, we're finally going to calculate the *exact* area beneath $y = x^2 + 1$ on the interval [0,3].

From now on, we're going to equate definite integrals with the area beneath a curve (technically speaking, the area between the function and the x-axis, remember?). Therefore, I can say that the area beneath $x^2 + 1$ on the interval [0,3] is equal to $\int_0^3 (x^2 + 1) \, dx$.

This new notation is read, "the integral of $x^2 + 1$, with respect to x, from 0 to 3." Unlike indefinite integrals, the solution to a definite integral, such as this one, is a number. That number is, in fact, the area beneath the curve. How in the world do you get that number, you ask? How about a warm welcome for the Fundamental Theorem?

The Fundamental Theorem of Calculus (part one): If $g(x)$ is the antiderivative of the continuous function $f(x)$, then $\int_a^b f(x) \, dx = g(b) - g(a)$.

CRITICAL POINT

> You will get a negative answer from a definite integral if the area in question is below the *x*-axis. Whereas the concept of "negative area" may not make sense to you, you automatically assign all area below the *x*-axis with a negative value.

In other words, to calculate the area beneath the curve *f(x)* on the interval [*a,b*], you must first integrate the function. Then, plug the upper bound (*b*) into the integral. From this value, subtract the result you get from plugging the lower bound (*a*) into the same integral. It's a brilliantly simple process, as powerful as it is elegant.

Example 4: Once and for all, find the *exact* area beneath the curve $f(x) = x^2 + 1$ on the interval [0,3] using the Fundamental Theorem of Calculus.

Solution: This problem asks you to evaluate the definite integral:

$$\int_0^3 \left(x^2 + 1\right) dx$$

CRITICAL POINT

> Here are two important properties of definite integrals:
>
> - $\int_a^a f(x) dx = 0$: If the upper and lower limits of integration are equal, the definite integral equals 0.
> - $\int_a^b f(x) dx = -\int_b^a f(x) dx$: You can swap the limits of integration if you like—just pop a negative sign out front.

Begin by integrating $x^2 + 1$ using the Power Rule for Integration. When you complete the integral, you no longer write the integration symbol, and you do not write "+ *C*." Instead, draw a vertical slash to the right of the integral, and copy the limits of integration onto it. This signifies that the integration portion of the problem is done:

$$\left(\frac{x^3}{3} + x\right)\Bigg|_0^3$$

Plug 3 into the function (for both *x*'s) and subtract 0 plugged into the function:

$$\left(\frac{3^3}{3} + 3\right) - \left(\frac{0^3}{3} + 0\right) = 9 + 3 = 12$$

YOU'VE GOT PROBLEMS

Problem 3: Calculate $\int_{\pi/2}^{3\pi/2} \cos x \, dx$. Explain what is meant by the answer.

Part Two: Derivatives and Integrals Are Opposites

I kind of spoiled this revelation for you already—I'm sorry. However, the second major conclusion of the Fundamental Theorem still holds some surprises. Let's check out the theorem first:

The Fundamental Theorem of Calculus (part two): If $f(x)$ is a continuous and differentiable function, $\frac{d}{dx}\left[\int_a^{f(x)} g(y) \, dy\right] = g\big(f(x)\big) \cdot f'(x)$, if a is a real number.

That looks unsightly. Here's what it means without all the gobbledygook. Let's say you're taking the derivative of a definite integral whose lower bound is a constant (i.e., just a number) and whose upper bound contains a variable. If you take the derivative of the entire integral with respect to the variable in the upper bound, the answer will be the function inside the integral sign (unintegrated), with the upper bound plugged in, multiplied by the derivative of the upper bound. This theorem looks, feels, and even smells complex, but it's not hard at all. Trust me on this one. All you have to do is learn the pattern.

Example 5: Evaluate $\frac{d}{dx}\left(\int_{\sin x}^3 t^2 \, dt\right)$.

Solution: You don't *have* to use the shortcut in part two of the Fundamental Theorem, but it makes things easier. Notice that the variable expression is in the lower (not the upper) bound, which is not allowed by the theorem. Therefore, you should swap them using a property of integrals I discussed earlier in the chapter. It says that flip-flopping the boundaries of an integral is fine, as long as you multiply the integral by –1.

$$\frac{d}{dx}\left(-\int_3^{\sin x} t^2 \, dt\right)$$

Because you are deriving with respect to x (and x is in the upper bound) and the lower bound is a constant, you are clear to apply the new theorem. All you do is plug the upper bound (sin x) into the function t^2 to get (sin x)^2, and multiply by the derivative of the upper bound (which will be cos x). Don't forget the negative, which stays out in front of everything:

$$-\sin^2 x \cdot \cos x$$

Here's a question for you: what if you forget this theorem? No problem—you can do Example 5 the long way, working from the inside out. Start with the integration problem and then take the derivative. You'll get the same thing. If you apply the Fundamental Theorem (part one) to the integral, you get:

$$\frac{d}{dx}\left(\frac{t^3}{3}\right)\Bigg|_{\sin x}^{3} = \frac{d}{dx}\left(9 - \frac{1}{3}\sin^3 x\right)$$

Take the derivative with respect to x to get $-\sin^2 x \cdot \cos x$. Don't forget to apply the Chain Rule when differentiating $\frac{1}{3}\sin^3 x$; that's where $\cos x$ comes from.

You don't always have to switch the boundaries and make the integral negative. Only do it if the constant appears in the upper boundary. What happens if both boundaries contain variables? If this is the case, you can't use the shortcut offered by the theorem and must resort to the long way.

YOU'VE GOT PROBLEMS

Problem 4: Evaluate $\frac{d}{dx}\left(\int_{1}^{\tan x} e^t \, dt\right)$ twice, once using the Fundamental Theorem of Calculus part one, and once using part two.

u-Substitution

At this point, you can't solve too many integration problems. You should have a handful of antiderivatives memorized (such as $\int \cos x \, dx = \sin x + C$ and $\int e^x \, dx = e^x + C$) and should have a pretty good grip on the Power Rule for Integration (meaning, for instance, you know that $\int x^7 \, dx = \frac{x^8}{8} + C$). However, what do you do if both of those techniques fail? You look, with hope glinting in your eyes, to a new method—u-substitution. You'll use u-substitution almost as much as the Power Rule for Integration—it's a calculus heavy hitter.

The key to u-substitution is finding a piece of the function whose derivative is also in the function. The derivative is allowed to be off by a coefficient, but otherwise must appear in the function itself.

Here are the steps you'll follow when u-substituting:

1. Look for a piece of the function whose derivative is also in the function. If you're not sure what to use, try the denominator or something being raised to a power in the function.

2. Set u equal to that piece of the function and take the derivative with respect to x.

3. Use your u and du expressions to replace parts of the original integral, and your new integral will be much easier to solve.

Example 6: Use u-substitution to find $\int \frac{\sin x}{\cos x}\,dx$ (i.e., prove that the integral of tangent is equal to $-\ln\left|\cos x\right| + C$).

Solution: Set u equal to a piece of the integral whose derivative is also in the integral. Because sine and cosine are both present (and the derivative of each is basically the other function), you could pick either one to be u, but remember the hint I gave you: if you're not sure which expression to choose, pick the denominator or something to a power. Therefore, set $u = \cos x$ and derive with respect to x to get $du = -\sin x\, dx$.

There's the $\sin x$ you expected. It, like $u = \cos x$, appears in the integral. Well, almost. In the original integral, $\sin x$ is positive, so multiply both sides of $du = -\sin x\, dx$ by -1 so that the sine functions match:

$$-du = \sin x\, dx$$

Now it's time to write the original integral with u's instead of x's. Instead of $\sin x$, the new numerator is $-du$ (because $-du = \sin x\, dx$). The new denominator is u (because $u = \cos x$).

$$\int \frac{-du}{u} = -\int \frac{du}{u}$$

Remember that the integral of $\frac{1}{x}$ is $\ln\left|x\right|$, so $-\int \frac{du}{u} = -\ln\left|u\right| + C$. The final step is to replace the u using your original u equation ($u = \cos x$) to get the final answer of $-\ln\left|\cos x\right| + C$.

The trickiest part of u-substitution is deciding what u should be. If your first choice doesn't work, don't sweat it. Try something else until it works out for you. It eventually will. The only way to get really good at this is to practice, practice, practice. Eventually, picking the correct u will become easier.

YOU'VE GOT PROBLEMS

Problem 5: Evaluate $\int_0^{\pi/4} \sec^2 x \cdot \tan x\, dx$. Hint: if you are performing u-substitution with a definite integral, you have to change the limits of integration as you substitute in the u and du statements. To change the limits, plug them each into the x slot of your u equation.

Tricky u-Substitution and Long Division

When we first discussed u-substitution, I made it a point to say that the derivative of u must appear in the problem. This is usually true, so I wasn't technically lying. There is a way to use u-substitution, even if it's not the most obvious choice.

Example 7: Find $\int \frac{2x-1}{x-2} \, dx$.

Solution: For grins, let's go ahead and try to find the antiderivative using *u*-substitution. Once again, remember our tip: if you're not sure what to set equal to *u*, try the denominator. Therefore, $u = x - 2$ and $du = dx$. If you make the appropriate substitutions back into the problem, you get:

$$\int \frac{2x-1}{u} \, du$$

To be honest, it doesn't look much better than the original, does it? Don't give up, though; we're not out of options. Go back to your *u* equation and solve it for *x* to get:

$$u = x - 2$$
$$x = u + 2$$

Now substitute that *x* value into the numerator of our integral, and suddenly everything is a little cheerier:

$$\int \frac{2(u+2)-1}{u} \, du$$
$$= \int \frac{2u+3}{u} \, du$$

At least all of our variables are the same now. That's a relief. Can you see where to go from here? This fraction is top-heavy, with lots of terms in the numerator but only one in the denominator, so we can use the separation method from last section to finish. What a happy coincidence that we just learned it!

$$\int \frac{2u}{u} \, du + \int \frac{3}{u} \, du$$
$$= 2\int \, du + 3\int \frac{1}{u} \, du$$
$$= 2u + 3\ln|u| + C$$
$$= 2(x-2) + 3\ln|x-2| + C$$
$$= 2x - 4 + 3\ln|x-2| + C$$
$$= 2x + 3\ln|x-2| + C$$

You may be wondering why the -4 vanished in the last step. Remember that *C* is some constant you don't know. If you subtract 4 from that, you'll get some other number (which is 4 less than the original mystery number). Since I *still* don't know the value for *C*, I just write it as *C* again, instead of writing $C - 4$.

There are alternatives when integrating fractions like these. In fact, you can begin a rational integral by applying long division; it helps to simplify the problem *if the numerator's degree is greater than or equal to the denominator's degree.* It works like a charm if the denominator is not a single term, as is the case with this example.

Because the degree of the numerator (1) is greater than or equal to the degree of the denominator (1), begin by dividing $2x - 1$ by $x - 2$:

$$\frac{2x-1}{x-2} \quad = \quad x-2\overline{)\begin{array}{c} 2 \\ 2x-1 \\ \underline{-2x+4} \\ 3 \end{array}} \quad = \quad 2+\frac{3}{x-2}$$

Therefore, you can rewrite the integral as $\int \left(2+\frac{3}{x-2}\right) dx$, and tricky u-substitution is no longer required. The solution will again be $2x + 3\ln|x-2| + C$.

YOU'VE GOT PROBLEMS

Problem 6: Find $\int \frac{2x+1}{2x-3}\, dx$ using tricky u-substitution or by using long division.

Technology Focus: Definite and Indefinite Integrals

Symbolic calculators like the TI-89 are very powerful integration machines. They work whether you're dealing with indefinite integrals (which have no bounds of integration) or definite integrals (which do).

In Example 1 from this chapter, you applied the Power Rule for Integration to determine that $\int \left(7x^3 + 6x^5\right) dx = \frac{7}{4}x^4 + x^6 + C$. To check the answer with your TI-89, you need to access the calculus tools in the $f3$ menu. Integration is option 2, as illustrated in Figure 15.1.

Figure 15.1

The integration tool is located just beneath the differentiate tool you used in Chapter 9.

As with differentiation, you follow the expression by a comma and then the variable that comes after the "*d*" in the problem. For example, this problem contains "*dx*" so you should type ",*X*" before you close the expression with a parenthesis (see Figure 15.2).

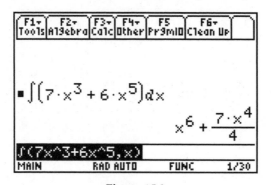

Figure 15.2
The solution verifies our solution to Example 1.

The calculator's solution looks a little different from our solution. For one thing, it lists the terms in order of exponent, from greatest to least. It also does not list the required "+ *C*" as part of the solution. Remembering that constant is up to you.

To compute a definite integral on your TI-89, enter the lower and upper bounds immediately following ",*X*" and separate them with commas. For example, in Example 4 you concluded that $\int_0^3 (x^2 + 1) dx = 12$. Check this answer by entering the expression in Figure 15.3.

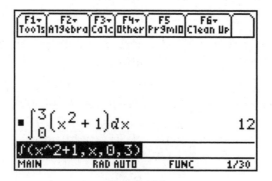

Figure 15.3
If you're faced with a definite integral, be sure to include the bounds.

Are you ready to be *really* impressed with a symbolic calculator? You can use them to check problems like Example 5, which are Fundamental Theorem of Calculus problems containing multiple variables (see Figure 15.4).

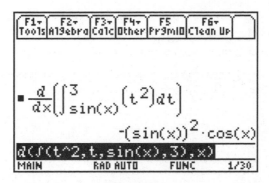

Figure 15.4

The TI-89 correctly computes the derivative, with respect to x, *of a definite integral written in terms of* t *that has a lower bound defined as a function of* x!

For those of you not wielding the humbling power of a symbolic calculator, hope is not lost. Your TI-84 may not be able to figure out definite integrals, but definite integrals are a snap. Press the MATH button and scroll down to "9:fnInt(" as illustrated in Figure 15.5.

Figure 15.5

You can't spell "fnInt" without "fun"! Never mind—you absolutely can.

If you have MathPrint enabled on your calculator, you are greeted with a fancy template into which you can plug your definite integral (see Figure 15.6).

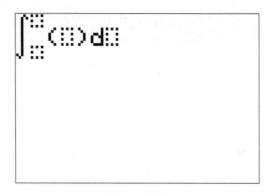

Figure 15.6

Insert your bounds, the expression you're integrating, and the variable you're integrating with respect to into the appropriate boxes.

When you do, you get the correct answer of 12, as illustrated in Figure 15.7.

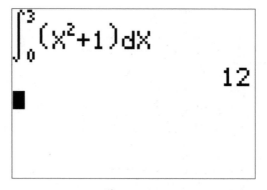

Figure 15.7

The right solution is definitely 12. Hooray for disappointing math puns!

If MathPrint is not enabled, you enter the expression exactly as you would on the TI-89 (see Figure 15.8).

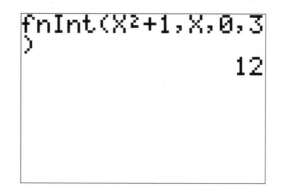

Figure 15.8
If you don't want to be bothered with arrowing between the boxes in the MathPrint template, you can compute the definite integral by typing this command.

The Least You Need to Know

- Integration, like differentiation, has a Power Rule of its own, in which you add 1 to the exponent and divide by the new exponent.

- Trigonometric functions have bizarre integrals, some of which are difficult to produce on your own; therefore, it's best to memorize them.

- The separation technique allows you to write an integral that's a sum or difference as a sum or difference of separate integrals.

- The two parts of the Fundamental Theorem of Calculus tell you how to evaluate a definite integral and give a shortcut for finding specific derivatives of integral expressions.

- *u*-substitution helps you integrate expressions that contain functions and their derivatives.

- Long division and tricky *u*-substitution are useful tools in your integration repertoire.

CHAPTER 16

Applications of the Fundamental Theorem

Once you learned how to find the slope of a tangent line
(a seemingly meaningless skill), it probably seemed as though
the applications for the derivative would never stop. You were
finding velocity and rates of change (both instantaneous and
average), calculating related variable rates, optimizing functions, determining extrema, and, all in all, bringing peace and
prosperity to the universe.

If you think that it's about time for applications of definite
integrals to start pouring in, you must be psychic. (Either
that or you read the table of contents.) For now, we'll look at
some of the most popular definite integral-related calculus
topics. We'll start by finding area bounded by two curves
(rather than one curve and the x-axis). We'll briefly backtrack
to topics we've already discussed, but we'll spice them up
a little with what we now know of integrals. Finally, we'll
look at definite integral functions, also called accumulation
functions.

In This Chapter

* Finding yet more curvy
  area

* Integration's Mean Value
  Theorem

* Position equations and
  distance traveled

* Functions defined by
  definite integrals

* Pretending you're Noah:
  finding arc length

Calculating Area Between Two Curves

This'll blow your mind. In fact, after you read it, you may question your very sanity. The thin ribbons of consciousness tying you to this mortal world may stretch and break, catapulting you into madness, or at least making you lose your appetite. Perhaps you should sit down before you continue.

> ✏️ **CRITICAL POINT**
>
> If you have functions containing y instead of functions containing x (i.e., $f(y) = y^2$), you can still calculate the area between the functions. However, instead of subtracting top minus bottom inside the integral, you subtract right minus left.

You've been calculating the area between curves *all along without even knowing it*. There, I said it. I hope you're okay.

If you want to calculate the area between two continuous curves, we'll call them $f(x)$ and $g(x)$, on the same x-interval $[a,b]$, here's what you do. Set up a definite integral as you did last chapter, with a and b as the lower and upper limits of integration, respectively. You'll stick either $f(x) - g(x)$ or $g(x) - f(x)$ inside the integral. To decide which one to use, you have to graph the functions—you should subtract the lower graph from the higher graph. For example, in Figure 16.1, $g(x)$ is below $f(x)$ on the interval $[a,b]$.

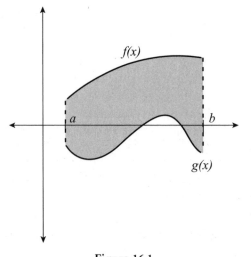

Figure 16.1

At least on the interval [a,b], the graph of f(x) is always higher than the graph of g(x).

Watch out! If you subtract the functions in the wrong order you'll get a negative answer, and you should *never* get a negative answer when finding the area between curves, even if some of that area falls below the *x*-axis.

What if the curves switch places? For example, look at the graph in Figure 16.2. To the left of *x* = *c*, *f*(*x*) is above *g*(*x*), but when *x* > *c*, the functions switch places and *g*(*x*) is on top.

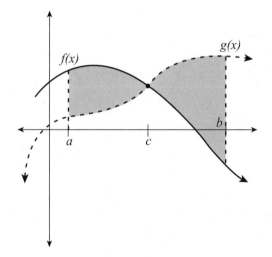

Figure 16.2
The graphs take turns on the top bunk—neither is above the other on the entire interval.

To find the shaded area, you'll have to use two separate definite integrals, one for the interval [*a*,*c*], when *f*(*x*) is on top, and one for [*c*,*b*], when *g*(*x*) is:

$$\int_a^c \left[f(x) - g(x) \right] dx + \int_c^b \left[g(x) - f(x) \right] dx$$

CRITICAL POINT

The reason we've technically been doing this all along is that we've always been finding the area between the curve and the *x*-axis, which has the equation *g*(*x*) = 0. Thus, if a function *f*(*x*) is above the *x*-axis on [*a*,*b*], the area beneath the two curves is $\int_0^3 \left[\left(x^2 + 1 \right) - 0 \right] dx$. That second equation has been invisible all this time.

Example 1: Calculate the area between the functions *f*(*x*) = sin 2*x* and *g*(*x*) = cos *x* on the interval $\left[\frac{3\pi}{2}, 2\pi \right]$.

Solution: These graphs play leapfrog all along the *x*-axis, but on the interval $\left[\frac{3\pi}{2}, 2\pi \right]$, *g*(*x*) is definitely above *f*(*x*) (see Figure 16.3).

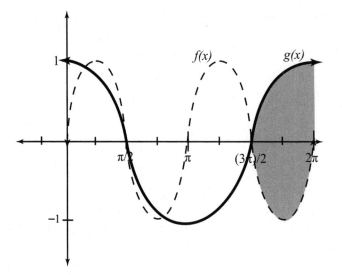

Figure 16.3
On the interval $\left[\frac{3\pi}{2}, 2\pi\right]$, g(x) = cos x rises above f(x) = sin 2x.

Therefore, the integral will contain $g(x) - f(x)$:

$$\int_{3\pi/2}^{2\pi} \left(\cos x - \sin 2x\right) dx$$

Split this up into separate integrals:

$$\int_{3\pi/2}^{2\pi} \cos x \, dx - \int_{3\pi/2}^{2\pi} \sin 2x \, dx$$

Calculating the first is fairly easy:

$$\int_{3\pi/2}^{2\pi} \cos x \, dx = \left(\sin x\right)\Big|_{3\pi/2}^{2\pi}$$

$$= \sin 2\pi - \sin \frac{3\pi}{2}$$

$$= 0 - \left(-1\right)$$

$$= 1$$

Use u-substitution to integrate sin $2x$, setting $u = 2x$:

$$\frac{1}{2}\int_{3\pi}^{4\pi} \sin u\, du = \frac{1}{2}\left(-\cos u\right)\Big|_{3\pi}^{4\pi}$$
$$= \frac{1}{2}\left[-\cos 4\pi - \left(-\cos 3\pi\right)\right]$$
$$= \frac{1}{2}\left[-(1)-\left(-(-1)\right)\right]$$
$$= \frac{1}{2}\left[-1-1\right]$$
$$= \frac{1}{2}\left(-2\right)$$
$$= -1$$

Don't forget to change your x boundaries into u boundaries when you u-substitute. For example, to get the new u boundary of 4π, plug the old x boundary of 2π into the u equation: $u = 2(4\pi) = 4\pi$. The final answer is the first integral minus the second:

$$\int_{3\pi/2}^{2\pi} g\left(x\right)dx - \int_{3\pi/2}^{2\pi} f\left(x\right)dx = 1-(-1) = 2$$

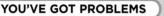

YOU'VE GOT PROBLEMS

Problem 1: Calculate the area between the curves $y = x^2$ and $y = x^3$ in the first quadrant.

The Mean Value Theorem for Integration

Think back to the original Mean Value Theorem from Chapter 13. It said that somewhere on an interval, the derivative was equal to the average rate of change for the whole interval. It turns out that integration has its own version of a Mean Value Theorem, but because integration involves area instead of rates of change, it's a bit different.

A Geometric Interpretation

In essence, the Mean Value Theorem for Integration states that at some point along an interval $[a,b]$, there exists a certain point $(c, f(c))$ between a and b (see Figure 16.4). If you draw a rectangle whose base is the interval $[a,b]$ and whose height is $f(c)$, the area of that rectangle will be exactly the area beneath the function on $[a,b]$.

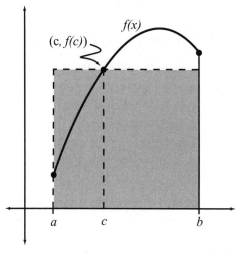

Figure 16.4

A visual representation of the Mean Value Theorem for Integration. The area of the shaded rectangle, whose height is f(c), *is exactly equal to* $\int_a^b f(x)dx$.

CRITICAL POINT

In the Mean Value Theorem for Integration, $(b - a)$ represents the length of the rectangle, because it is the length of the interval $[a,b]$. The height of the rectangle is, as we've already discussed, $f(c)$. There may be more than one such c in the interval that satisfies the Mean Value Theorem for Integration, but there must be at least one.

The Mean Value Theorem for Integration: If a function $f(x)$ is continuous on the interval $[a,b]$, then there exists c, $a \le c \le b$, such that $(b-a) \cdot f(c) = \int_a^b f(x)dx$.

This Mean Value Theorem, like its predecessor, is only an existence theorem. It guarantees that the value $x = c$ and the corresponding key height $f(c)$ exist. You may wonder why it's so important that a curvy graph and a plain old rectangle must always share the same area. We'll get to that after the next example.

Example 2: Find the value $f(c)$ guaranteed by the Mean Value Theorem for Integration for the function $f(x) = x^3 - 4x^2 + 3x + 4$ on the interval $[1,4]$.

Solution: The Mean Value Theorem for Integration states that there is a c between a and b so that $(b-a) \cdot f(c) = \int_a^b f(x)dx$. You know everything except what c is, but that's okay.

Plug in everything you know:

$$(4-1)\cdot f(c)=\int_1^4\left(x^3-4x^2+3x+4\right)dx$$

CRITICAL POINT

If Example 2 had asked you to find the c-value rather than the value of $f(c)$, you'd still follow the same steps. At the end, however, you'd plug the point

$$3f(c)=\frac{57}{4}$$
$$f(c)=\frac{57}{4}\cdot\frac{1}{3}$$
$$f(c)=\frac{57}{12}$$
$$f(c)=\frac{19}{4}$$
$$f(c)=4.75 \text{ into } f(x) \text{ and solve for } c.$$

After the quick subtraction problem on the left (and the slightly lengthier definite integral on the right), you should get this:

$$\frac{57}{4}$$

This means that the area beneath the curve $f(x) = x^3 - 4x^2 + 3x + 4$ on the interval $[1,4]$ (which is $\frac{19}{4}$) is equal to the area of the rectangle whose length is the same as the interval's length (3) and whose height is $\left(c,\frac{19}{4}\right)$.

YOU'VE GOT PROBLEMS

Problem 2: Find the value $f(c)$ guaranteed by the Mean Value Theorem for Integration on the function $f(c)=\dfrac{\int_a^b f(x)dx}{b-a}$ on the interval [1,100].

The Average Value Theorem

The value $f(c)$ that you found in both Example 2 and Problem 2 has a special name. It is called the *average value* of the function. If you take the Mean Value Theorem for Integration and divide both sides of it by $(b - a)$, you'll get the equation for average value:

$$f(x)=\frac{\ln x}{x}$$

📖 **DEFINITION**

The **average value** of a function is the value $f(c)$ guaranteed by the Mean Value Theorem for Integration (the height of the rectangle of equal area). The average value is found via the equation

$$\int_0^{3.78586}\left(-15.5t^3 + 86.25t^2 - 117.25t + 48.75\right)dx - \int_{3.78586}^{4}\left(-15.5t^3 + 86.25t^2 - 117.25t + 48.75\right)dx\,.$$

This is simply a different form of our previous equation, so it doesn't warrant much more attention. However, some textbooks completely skip over the Mean Value Theorem for Integration and go right to this, which they call the Average Value Theorem. They might word Problem 2 earlier as follows: "Find the *average value* of $f\left(c\right) = \frac{\int_a^b f\left(x\right)dx}{b-a}$ on the interval $[1,100]$." You'd solve the problem the exact same way (see Figure 16.5).

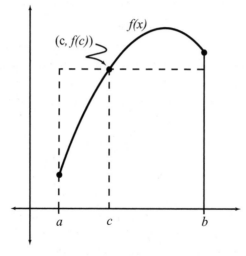

Figure 16.5

Here's the diagram for the Mean Value theorem for Integration once more. The height of the denoted line is the function's average value. Although the function dips below and shoots above f(x), that's how high f(x) is on average.

Here's one way to think of the relationship between the theorems. Most functions twist and turn throughout their domains. If you could "level out" a function by filling in its valleys and flattening out its peaks until the function was a horizontal line, the height of that line (i.e., its y-value) would be the average value for that function.

Finding Distance Traveled

Definite integrals also play well with position and velocity functions. Remember that derivatives measure a rate of change. Well, it turns out that definite integrals of rate of change functions measure accumulated change. For example, if you are given a function that represents the rate of sales of the new must-have toy, the Super Fantastic Hula Hoop, then the definite integral gives you the actual number of hula hoops sold.

Most often, however, math teachers like to explore this property of integrals as it applies to motion. Specifically, the definite integral of the velocity function of an object gives you the total displacement of the object. A word of caution: you will most often be asked to find the total *distance* traveled by the object—not the total displacement. To calculate the total distance, you'll first have to determine where the object changes direction (using a wiggle graph) and then integrate the velocity separately on every interval that direction changes.

Here's the difference between total distance traveled and total displacement. Let's say at any hour t, I want to know (in miles) how far I am away from my favorite bright orange 1970s-style easy chair that my wife hates. My initial position (i.e., $t = 0$ hours) is in the chair, so $s(0) = 0$. Two hours later, I am at work, 50 miles away from my chair, so $s(2) = 50$ miles. Once my workday and commute home are complete, I am back home in the chair, and $s(12) = 0$. I have traveled a total distance of 100 miles, counting my travel away from the chair and back again. However, my displacement is 0. Displacement is the total change in position counting only the beginning and ending position; if the object in question changes direction any time during that interval of time, displacement does not correctly reflect the total distance traveled.

Example 3: In the book *The Fellowship of the Ring* by J.R.R. Tolkien, a young hobbit named Frodo embarks on an epic, exciting, and hairy-footed adventure to destroy the One Ring in the fires of Mount Doom. Based on a little estimation and the book *Journeys of Frodo: An Atlas of J.R.R. Tolkien's The Lord of the Rings* by Barbara Strachey, I have designed an equation modeling Frodo's journey. During the first four days of his journey (from Hobbiton to the home of Tom Bombadil), his velocity (in miles per day) is given by this equation:

$$v(t) = -15.5t^3 + 86.25t^2 - 117.25t + 48.75$$

For example, $v(2)$ gives his approximate velocity at the exact end of the second day. Find the total distance Frodo travels from $t = 0$ to $t = 4$.

Solution: Because you want to find the total distance traveled, you need to determine if Frodo changed direction at any point, and actually started to wander toward Hobbiton rather than away. This is not necessarily caused by poor hobbit navigation, but perhaps by hindrances such as the old forest, getting caught in trees, etc. To see if his direction changed, create a wiggle graph for velocity (see Figure 16.6).

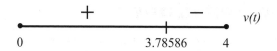

Figure 16.6
The hobbits have a pretty good sense of direction; in fact, they are heading farther and farther away from Hobbiton until just before the end of the fourth day.

Integrate the velocity equation separately, on both of these intervals. Because they are heading slightly backward (i.e., toward their beginning point) on the interval (3.78586,4), that definite integral will be negative. However, because it still represents the distance the hobbits are traveling, you don't want it to be subtracted from your answer, so turn it into its opposite by multiplying it by –1. You should do this for any negative pieces of your wiggle graph in this type of problem. Therefore, the distance traveled is:

$$\ln x = \int_1^x \frac{1}{t} \, dt$$

Although the numbers are darn ugly, the premise is very simple. I'll leave the figuring up to you. You should get 108.298 for the first interval (distance away from Hobbiton) and –(–3.298) for the second, which is the small distance back toward Hobbiton; the sum is 111.596 miles.

Right about now, you're seeing that the numbers in this problem are not easy whole numbers. They rarely turn out to be so in real-world (or Middle Earth) examples, so calculators are a necessary tool. There are those who would have me burned at the mathematical stake for suggesting such a thing. In fact, I was once yelled at fiercely by the lunch ladies in the high school cafeteria where I worked for suggesting that you should use a calculator to check your answers when converting fractions to decimals. I've never received fewer tater tots than I did that day. Lunch ladies can be so bitter.

YOU'VE GOT PROBLEMS

Problem 3: When satellites circle closely around a planet or moon, the gravitational field surrounding the celestial body both increases the satellite's velocity and changes its direction in an orbital move called a "slingshot." (As you may know from the movie *Apollo 13*, Tom Hanks and his crew executed a slingshot maneuver around the moon to hurl themselves back toward Earth.) Let's say that a ship executing this maneuver has position equation $s(t) = t^3 - 2t^2 - 4t + 12$, where t is in hours and $s(t)$ represents thousands of miles from Earth. What is the total distance traveled by the craft during the first five hours?

Accumulation Functions

Before we close out this chapter and make it a fond memory, let's talk about accumulation functions. You'll probably see a few of them lurking around contemporary calculus classes, as they are now "in" because of the advent of calculus reform. An *accumulation function* is a definite integral with a variable expression in one or more of its limits of integration. They are called accumulation functions because they get their value by accumulating area beneath curves, as do all definite integrals.

> **DEFINITION**
>
> An **accumulation function** is a function defined by a definite integral; the function will have a variable in one or both of its limits of integration.

The most famous accumulation function is the natural logarithmic function: $< y = \frac{1}{t}$. The natural log function gets its value by accumulating area under the simple curve $y = \frac{1}{t}$! For example, the value of ln 5, which always seemed so alien to me (where the heck do you get 1.60944?) is equal to the area beneath $f(x) = \int_{2}^{x}(t-4)dt$ on the interval $[1,5]$.

Practically speaking, you should be able to evaluate and differentiate accumulation functions, so let's get to it. Believe it or not, evaluating accumulation functions is just as easy as evaluating any other function—just plug in the correct x-value. Once you plug in the value, you'll apply the Fundamental Theorem to the resulting definite integral.

Example 4: Given the accumulation function $f(4) = \int_{2}^{4}(t-4)dt$, complete the following tasks:

(a) Calculate $f(4)$.

(b) Determine $f'(x)$.

Solution:

(a) Plug 4 into x, not t, because $f(x)$ is a function of x. In other words, x becomes the upper limit of integration:

$$f(4) = \left(\frac{t^2}{2} - 4t\right)\Big|_{2}^{4}$$

$$= \left(\frac{16}{2} - 16\right) - \left(\frac{4}{2} - 8\right)$$

$$= (8 - 16) - (2 - 8)$$

$$= -8 - (-6)$$

$$= -2$$

Now you can integrate like normal:

$$g(x) = \int_{-\pi}^{x/2} \cos 2t \, dt$$

(b)   To find the derivative of an accumulation function, look no further than part two of the Fundamental Theorem. In this case, $f'(x) = x - 4$. Just plug the top bound into t and multiply by its derivative (which is 1 in this case). Pretty easy, eh? You already knew how to tackle these problems, even before they showed up. Kudos!

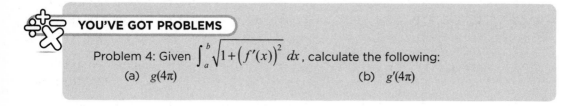

YOU'VE GOT PROBLEMS

Problem 4: Given $\int_{a}^{b} \sqrt{1 + \left(f'(x)\right)^2} \, dx$, calculate the following:
   (a) $g(4\pi)$ (b) $g'(4\pi)$

Arc Length

At this point, you can do all kinds of crazy math calculating. Geometry told you how to find weird areas, and calculus took that one step further. The kinds of areas you can calculate now would have boggled your mind back in your days of geometric innocence. However, it remains a math skill that has visible and understandable application, even to those who don't know the difference between calculus and a tuna sandwich. Now, let's add to your list of skills the ability to find lengths of curves. By the time you're done, you'll even be able to prove (finally) that the circumference of a circle really is 2π.

Rectangular Equations

The term "rectangular equations" really means "plain old functions." Mathematicians use the term for the obvious reason that it takes less time to say (mathematicians are busy people). It turns out that finding the length of a curve (on some x interval) is as easy as calculating a definite integral. In fact, the length of a continuous function $f(x)$ on the interval $[a,b]$ is equal to $g(x) = \sqrt{x}$. In other words, find the derivative of the function, square it, add 1, and integrate the square root of the result over the correct interval.

Example 5: Find the length of the function $g'(x) = \frac{1}{2\sqrt{x}}$ between points (1,1) and (16,4) on its graph.

Solution: Use the Power Rule to find the derivative of $g(x) = x^{1/2}$, and you get

$\int_1^{16} \sqrt{1 + \left(\frac{1}{2\sqrt{x}}\right)^2}\, dx = \int_1^{16} \sqrt{1 + \frac{1}{4x}}\, dx$. All you do now is plug into the arc length formula:

$$\int_a^b \sqrt{\left(\frac{dx}{dt}\right)^2 + \left(\frac{dy}{dt}\right)^2}\, dt$$

The integration problem that results is not simple at all. For our purposes, it is enough to know and apply the formula, not to struggle through the integral itself. You'll find that many (if not most) arc length integrals will end up complicated and require somewhat advanced methods to integrate. We will, however, satisfy ourselves with a computer- or calculator-assisted solution—they have no problem with complex definite integrals. The final answer is approximately 15.3397.

Don't feel like you're cheating by using a calculating tool rather than solving this problem by hand. You'd have to know just about every integration technique there is to find the arc lengths of even very simple functions.

> **YOU'VE GOT PROBLEMS**
>
> Problem 5: Which function is longer on the interval [0,2], $f(x) = x^2$ or $g(x) = x^3$? Find the length of each and compare.

Parametric Equations

We haven't mentioned parametric equations for a while—they've been lurking in the shadows, but now they get to come out and play. There are numerous similarities in the formula for parametric equation arc length and rectangular arc length. Both are definite integrals, both involve a sum of two terms beneath a radical, and both involve finding the derivative of the original equation.

The arc length of a curve defined parametrically is found with this definite integral:

$$0 \le \theta \le 2\pi$$

In other words, find the derivatives of the x and y equations, square them both, add them together, take the square root, and integrate the whole mess. Note that a and b are limiting values of the parameter t this time—not x boundaries.

Example 6: The parametric representation of a circle with radius 1 (centered at the origin) is $x = \cos\theta, y = \sin\theta$. Prove that the circumference of a circle really is 2π by calculating the arc length of the parametric curve on $\frac{dx}{d\theta} = -\sin\theta$ and $\frac{dy}{d\theta} = \cos\theta$.

KELLEY'S CAUTIONS

Don't get confused because the parameter in Example 6 is not t. The formula for arc length with a parameter of θ is exactly the same—it just has θ's instead of t's in the formula.

Solution: Start by finding the derivatives of the x and y equations with respect to θ (it's easy):

$$\int_0^{2\pi} \sqrt{(-\sin\theta)^2 + (\cos\theta)^2}\ d\theta$$

$$= \int_0^{2\pi} \sqrt{1}\ d\theta$$

$$= (\theta)\ \Big|_0^{2\pi}$$

$$= 2\pi - 0 = 2\pi$$

Now, plug those values into the parametric arc-length formula and simplify using the Mama theorem (review Chapter 4 if you don't know what the heck I mean by that):

$$\int_a^b f(x)\,dx$$

YOU'VE GOT PROBLEMS

Problem 6: Find the arc length of the parametric curve defined by the equations $x = t + 1$, $y = t^2 - 3$ on the t-interval [1,3].

The Least You Need to Know

- To find the area between two curves, integrate the curve on top subtracted by the curve below it on the proper interval.

- The average value of a function, $f(x)$, over an interval, $[a,b]$, is found by dividing the definite integral of that function, $\int_a^b f(x)\,dx$, by the length of the interval itself, $b - a$.

- To calculate the distance traveled by an object, calculate the definite integral of its velocity function separately for each period of time it changes direction.

- Accumulation functions get their value by gathering area under a curve; they are defined by definite integrals with variables in one or more of their limits of integration.

- You can find the arc length of rectangular or parametric curves via similar definite integral formulas.

PART 5

Differential Equations and More

The final leg of your Calculus I journey ends here, in the strange sea of differential equations, a churning whirlpool of complicated mathematics. Not to worry—as a calculus student, you won't dive headfirst into the whirlpool so much as just dip your big toe in.

Why end with differential equations? They are an extension of everything you've worked on so far. They are literally equations containing derivatives that you will solve using integration. All your preparation is about to pay off!

Finally, it's time to put away your notes, stick your backpack under your chair, and take the final exam to find out how much you've learned. Eyes on your own paper!

CHAPTER 17

Differential Equations

Most calculus courses contain some discussion of differential equations, but that discussion is extremely limited to the basics. Most math majors will tell you that they had to suffer through an entire course on solving differential equations at some point in their math career. This is because differential equations are extremely useful in modeling real-life scenarios, and are used extensively by scientists.

A differential equation is nothing more than an equation containing a derivative. In fact, you have created more than your fair share of differential equations simply by finding derivatives of functions in the first half of the book. In this chapter, you'll begin with the differential equation (i.e., the derivative) and work your way backward to the original equation. Sound familiar? Basically you're just going to be applying integration methods, as you have for numerous chapters now.

However, solving differential equations is not the same thing as integrating. There are lots of complicated differential equations (that we won't be exploring). Luckily, the most popular differential equation application in beginning calculus (exponential growth and decay) requires you to use a very simple solution technique called separation of variables. Let's start there.

In This Chapter

- What are differential equations?
- Separation of variables
- Initial conditions and differential equations
- Modeling exponential growth and decay

Separation of Variables

If a *differential equation* is nothing more than an equation containing a derivative, and solving a differential equation basically means finding the antiderivative, then what's so hard about solving differential equations, and why does it get treated as a separate topic? The reason is that differential equations are usually not as straightforward as this one:

$$\frac{dy}{dx} = 3\cos x + 1$$

Clearly, the solution to this differential equation is $y = 3\sin x + x + C$. All you have to do is integrate both sides of the equation. Most differential equations are all twisted up and knotted together with variables all over the place, like this:

$$\frac{dy}{e^{2x}} = xy\,dx$$

It looks like someone chewed up a whole bunch of equations and spat them out in random order (which is both puzzling and unappetizing). In order to solve this differential equation, you'll have to separate the variables. In other words, move all the y's to the left side of the equation and all the x's to the right side. Once that is done, you'll be able to integrate both sides of the equation separately. This process, appropriately called *separation of variables*, solves any basic differential equations you'll encounter.

> **DEFINITION**
>
> **Differential equations** are just equations that contain a derivative. Most basic differential equations can be solved using a method called **separation of variables,** in which you move different variables to opposite sides of the equation so that you can integrate both sides of the equation separately.

Example 1: Solve the differential equation $\frac{dy}{dx} = ky$, where k is a constant.

Solution: You need to move y to the left side of the equation and move dx to the right side. Because k is a constant, it's not clear whether or not you should move it. As a rule of thumb, move all constants to the right side of the equation. Your goal is to solve for y, so you don't want any non-y things on the left side of the equation. Start by moving the y, so divide both sides of the equation by y to get:

$$\frac{dy}{y \cdot dx} = k$$

Now shoot that dx to the right side of the equation by multiplying both sides by dx:

$$\frac{dy}{y} = k\,dx$$

At this point, you can integrate both sides of the equation. Because k is a constant, its antiderivative is kx, just as the antiderivative of 5 would be $5x$:

$$\int \frac{dy}{y} = \int k\,dx$$
$$\ln|y| = kx + C$$

You're not quite done yet. Your final answer to a differential equation should be solved for y. To cancel out the natural log function and accomplish this, you have to use its inverse function, e^x, like this: $e^{\ln|y|} = e^{kx+C}$. (Drop the absolute value signs around the y now—they were only needed because the domain of the natural log function is only positive numbers. As the natural log disappears, let the absolute value bars go with it.)

In other words, rewrite the equation so that both sides are the powers of the natural exponential function. This gives you $y = e^{kx+C}$, because e^x and $\ln x$ are inverse functions, and as such, $e^{\ln x} = \ln(e^x) = x$. You could stop here, but go just one step further. Remember the basic exponential rule that said $x^a \cdot x^b = x^{a+b}$? The preceding equation looks like x^{a+b}, so you can break it up into $x^a \cdot x^b$:

$$y = e^{kx} \cdot e^C$$

Almost done. I promise. Because you have no idea what value C has, you have no idea what e^C will be. You know it'll be some number, but you have no idea what number that is. As you've done in the past, rewrite e^C as C, signifying that even though it's not the same value as the original C, it's still some number you don't know: $y = Ce^{kx}$.

That's the solution to the differential equation. It took you a while to get here, but this is a very important equation, and you'll need it in a few pages.

YOU'VE GOT PROBLEMS

Problem 1: Solve the differential equation: $\left(x^2 - 1\right)dy = \frac{x\,dx}{\cos y}$.

Types of Solutions

Just like integrals, solutions to differential equations come in two forms: with and without a "+ C" term. Definite integrals had no such term, because their final answers were numbers rather than equations. Whereas the solution to a differential equation will always come in delicious, equation form (with candy-shaped marshmallows), in some cases, you'll be able to determine exactly what the value of C should be, so you can provide a more specific answer.

Family of Solutions

If you are only given a differential equation, you can only get a general solution. Example 1 and Problem 1 are two such instances. Remember, integration cannot usually tell you exactly what a function's antiderivative is, because any functions differing only by a constant will have the same derivative.

The solution to a differential equation containing a "+ C" term is actually a *family of solutions,* because it technically represents an infinite number of possible solutions to the differential equation. Think about the differential equation $\frac{dy}{dx} = 2x + 7$. If you use the separation-of-variables technique, you get a solution of $y = x^2 + 7x + C$. You can plug in any real-number value for C, and the result is a solution to the original differential equation.

For example, $y = x^2 + 7x + 5$, $y = x^2 + 7x - \frac{105}{13}$, and $y = x^2 + 7x + 4\pi$ all have a derivative of $2x +$ 7. These three (plus an infinite number of other equations) make up the family of solutions. The members of a family of solutions have nearly identical graphs, differing only in their vertical position along the y-axis.

> **DEFINITION**
>
> Any mathematical solution containing "+ C" is called a **family of solutions,** because it doesn't give one specific answer. It compactly describes an infinite number of solutions, each differing only by a constant.

Knowing a family of solutions is sometimes not enough. Differential equations are often used as mathematical models to illustrate real-life examples and situations. In such cases, you'll need to be able to find specific solutions to differential equations, but to do so you'll need a little more information up front.

Example 2: Graph the family of solutions for the differential equation $\frac{dy}{dx} = 3x^2 + 2x - 6$ when $C = -4, -2, 0, 2,$ and 4.

Solution: To solve the differential equation, begin by separating the variables. In other words, multiply both sides of the equation by dx to isolate the y-variables on the left side of the equal sign and the x-variables on the right side:

$$dy = (3x^2 + 2x - 6)\, dx$$

Now integrate both sides of the equation separately.

$$\int dy = \int \left(3x^2 + 2x - 6\right) dx$$
$$y = \frac{3x^3}{3} + \frac{2x^2}{2} - 6x + C$$
$$y = x^3 + x^2 - 6x + C$$

This is a family of solutions, because substituting any real number C into the equation creates a new solution whose derivative (with respect to x) is the original differential equation. In this problem, you are asked to graph specific members of the family of solutions. Plug each given value of C into the solution to create five unique solutions:

$$y = x^3 + x^2 - 6x - 4$$
$$y = x^3 + x^2 - 6x - 2$$
$$y = x^3 + x^2 - 6x$$
$$y = x^3 + x^2 - 6x + 2$$
$$y = x^3 + x^2 - 6x + 4$$

The only difference in each equation is the constant, the value at which the graph crosses the y-axis when $x = 0$. Therefore, the graphs are all vertical translations (or shifts) of each other, as illustrated in Figure 17.1.

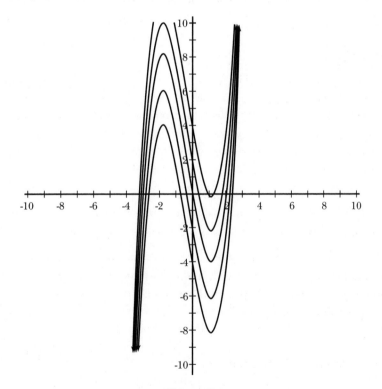

Figure 17.1
The family of solutions to the differential equation when C = −4, −2, 0, 2, and 4.

Specific Solutions

To determine exactly what C equals for any differential equation solution, you'll need to know at least one coordinate pair of the differential equation's antiderivative. With that information, you can plug in the (x,y) pair and solve for C. To explain what I mean, I have thrown together a little example for those game show fans out there.

I don't know what it is that makes people (by which I mean my wife and me) so excited to watch other people agonize about winning dishwashers by throwing comically oversized dice, but it doesn't stop us from watching game shows. However, the sudden trend in these programs isn't playing silly games for prizes anymore; instead, they subject the contestants to peril in order to win vast sums of money.

Example 3: A new television game show in the works already is making quite a stir. *Terminal Velocity* will suspend contestants by their ankles on a bungee cord. Producers are still working out the details, but one of the show's features will be dropping the participants from the ceiling and allowing them to repeatedly lurch their way toward the studio floor as the length of the bungee slowly increases. Also, the audience will throw things at them (like small rocks or maybe piranhas if ratings begin to sag).

Suppose that the velocity of a contestant (in ft/sec), for the first 10 seconds of her fall, is given by $\frac{ds}{dt} = -80\sin(2t) - 4$. If the initial position of the doomed individual is 115 feet off the ground, find her position equation.

Solution: You are given a differential equation representing velocity. The solution to the differential equation will then be the antiderivative of velocity, position. The problem also tells you that the initial position is 115 feet high. This means that the contestant's position at time equals 0 is 115, so $s(0) = 115$. You'll use that in a second to find C, but first things first—you need to apply separation of variables to solve the differential equation:

$$\int ds = \int \left[-80\sin(2t) - 4 \right] dt$$
$$s(t) = 40\cos(2t) - 4t + C$$

There you have it—the position equation. Remember that you should get 115 if you plug in 0 for t; make that substitution, and you can find C easily:

$$115 = 40\cos(2 \cdot 0) - 4 \cdot 0 + C$$
$$115 = 40\cos(0) + C$$
$$115 = 40 \cdot 1 + C$$
$$75 = C$$

Therefore, the exact position equation is $s(t) = 40\cos(2t) - 4t + 75$.

YOU'VE GOT PROBLEMS

Problem 3: A particle moves horizontally back and forth along the x-axis according to some position equation $s(t)$; the particle's acceleration (in ft/sec^2) is described accurately by the equation $a(t) = 2t + 5 - \sin t$. If you know that the particle has an initial velocity of –2 ft/sec and an initial position of 5 feet, find $v(t)$ (the particle's velocity) and $s(t)$ (the particle's position).

Exponential Growth and Decay

Most people have an intuitive understanding of what it means to exhibit *exponential growth*. Basically, it means that things are increasing in an out-of-control way, like a virus in a horror movie. One infected person spreads the illness to another person, then those two spread it to other people. Two infected people become four, four become eight, eight become sixteen, until it's an epidemic and Jackie Chan has to come in to save the day, possibly with karate kicks.

Truth be told, actual exponential growth doesn't happen a lot. An exponential growth model assumes that there is an infinite amount of resources from which to draw. In our epidemic example, the rate of increase of the illness cannot go on uninhibited, because eventually, everyone will already be sick. To get around such restrictions, many problems involving exponential growth and decay deal with exciting things like bacterial growth. Bacteria are small, so it takes them much longer to conquer the world (thanks in no small part, once again, to Jackie Chan).

DEFINITION

Exponential growth occurs when the rate of change of a population is proportional to the population itself. In other words, the bigger the population, the larger it grows. **Logistic growth** begins almost exponentially but eventually grows more slowly and stops, as the population reaches some limiting value.

Notice that in Figure 17.2, the *logistic growth* curve changes concavity (from concave up to concave down) about midway through the interval. This change in concavity indicates the point at which growth begins to slow.

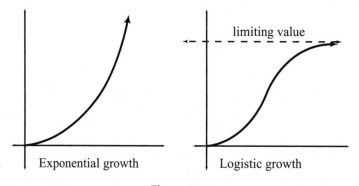

Figure 17.2
Two kinds of growth. Neither explains that weird mole on your neck.

A more realistic example of growth and decay is logistic growth. In this model, growth begins quickly (it basically looks exponential at first) and then slows as it reaches some limiting factor (as our virus could only spread to so many people before everyone was already dead—isn't that a pleasant thought?). Although it is not beyond our abilities to examine logistic growth, it is by far more complicated to understand and model, so we'll stick with exponential growth.

Mathematically, exponential growth is pretty neat. We say that a population exhibits exponential growth if its rate of change is directly proportional to the population itself. Thus, a population P grows (or decays) exponentially if $\frac{dP}{dt}$ and P are in proportion to each other.

So how fast something grows or decays is based on how much of it there is. Without getting into a lot of detail (too late for that, isn't it?), we can say that these two things are in proportion when they're equal to each other, if one of the terms is multiplied by a constant (for instance, one thing is two times or five times as big as the other).

$$\frac{dP}{dt} = k \cdot P$$

Recognize that? It's Example 1! Because you already solved this differential equation earlier in the chapter, you know that a population showing exponential growth has this equation:

$$y = Ne^{kt}$$

(I know I used C as the constant before, but I like having the N there better, because when you read the formula, it looks like the word *naked,* and I am immature enough to think that's pretty funny.) In this formula, N represents the beginning or initial population, k is a constant of proportionality, and *t* stands for time. (The *e* is a constant—Euler's number, which you've undoubtedly seen lurking about in your precalculus work. Most calculators, even scientific ones, have a button for Euler's number, so you don't have to memorize it.) The *y* represents the total population after time *t* has passed.

Your first step in the majority of exponential growth and decay problems is to find *k*, because you will almost never be able to determine what *k* is based on the problem. Don't even try to guess *k*—it's rarely, if ever, obvious. For example, if your population increased in the familiar sequence 2, 4, 8, 16, 32, etc., you might be tempted to think that *k* = 2 because the population constantly doubles, but instead, $k \approx 0.693147$, which is actually ln 2.

✎ CRITICAL POINT

You use the same formula for both exponential growth and decay. The only difference in the two is that *k* will turn out to be negative in decay problems and positive in growth problems. In both cases, *N* represents the initial value.

Example 4: Even after the movie *Pay It Forward* came out, the movement promoted by the film never really caught on. Its premise was that you should do a big favor for three different people, something they couldn't accomplish on their own. In turn, they would provide favors for three other people, and so on. Unfortunately, a new movement called Punch It Forward is catching on instead. It's the same premise, but with punching instead of favors. On the first day of Punch It Forward, 19 people are involved in the movement. After 10 days, 193 people are involved. How many people will be involved 30 days after Punch It Forward begins, assuming that exponential growth is exhibited during that time?

Solution: Use the exponential growth and decay formula $y = Ne^{kt}$. *N* represents the initial population (19). You know that after 10 days have elapsed, the new population is 193. Therefore, when $t = 10$, $y = 193$. Plug all these values in and solve for *k*:

$$y = Ne^{kt}$$
$$193 = 19e^{10k}$$
$$\frac{193}{19} = e^{10k}$$
$$\ln\left(\frac{193}{19}\right) = 10k$$
$$\frac{\ln(193/19)}{10} = k$$
$$0.231825 \approx k$$

Therefore, the exponential growth model is $y = 19e^{0.231825t}$. To determine the population after the first 30 days, plug in 30 for t:

$$y = 19e^{0.231825(30)}$$
$$y \approx 19{,}914.2$$

Approximately 19,914 people have been inducted (and possibly indicted) into the Punch It Forward society after only one month. It's a brave new world, my friend.

Example 5: According to Sir Isaac Newton (the very same cofounder of calculus discussed in Chapter 1), the rate at which an object cools is directly proportional to the temperature of its environment. This property is known as Newton's Law of Cooling. Now that I have shared this tasty nugget of information with you, how about a couple of questions to help you digest it?

(a) Assume that an object cools at a rate of $\frac{dT}{dt}$, where T is the temperature of the object after time t has elapsed. Furthermore, the object's environment has temperature T_E and the constant of proportionality described by Newton is equal to k. Create a differential equation based on Newton's Law of Cooling and solve it for T. Then, identify the specific solution stating that the original temperature of the object (when $t = 0$ and no time has passed) equals T_0.

(b) According to scientists, the optimum temperature for tea is 140°F. You leave a cup of tea at this temperature in a 72°F room, and exactly 17.5 minutes later it reaches 113°F, the minimum temperature at which scientists deem the tea still drinkable. At approximately what time was the tea's temperature 125°F?

Solution:

(a) The first part of this problem requires you to generate Newton's Law of Cooling. It states that the rate $\frac{dT}{dt}$ at which an object's temperature T changes is proportional to the difference between the object's temperature T and the temperature of the environment T_E.

$$\frac{dT}{dt} = k\left(T - T_E\right)$$

Recall that if two values are proportional, it means you can multiply one of those values by a constant (in this case k) to get the other value. Separate the variables, dividing both sides of the equation by $T - T_E$ and multiplying both sides by dt:

$$\frac{dT}{T - T_E} = k\,dt$$

Integrate both sides of the equation:

$$\int \frac{dT}{T - T_E} = \int k\, dt$$

Note that T_E and k are constants, so treat them as you would any other number. That means $\int k\, dt = k \cdot t + C$. Use u-substitution to integrate the left side of the equation. If $u = T - T_E$, then $du = dt$ and $\int \frac{du}{u} = \ln|u| + C = \ln|T - T_E| + C$. Set these two new integrals equal to each other as described in the above differential equation:

$$\ln|T - T_E| = kt + C$$

The absolute value bars are unnecessary. The object will eventually cool to room temperature (T_E), but it will not get colder than that. Because $T \geq T_E$, you can conclude that $T - T_E \geq 0$.

$$\ln (T - T_E) = kt + C$$

Don't lose sight of your goal, to solve this equation for T. Exponentiate both sides of the equation, making them powers of e, to eliminate the natural logarithm.

$$e^{\ln(T - T_E)} = e^{kt} e^C$$

Recall that e^C is another unknown constant, which can be represented once again by C.

$$T - T_E = e^{kt} \cdot C$$
$$T - T_E = Ce^{kt}$$
$$T = T_E + Ce^{kt}$$

Whew! That's a lot of work but finally you have an equation solved for T. This is a family of solutions, and to complete this problem, you need to identify the specific solution for which $T = T_0$ when $t = 0$. Your goal is to figure out what C should equal:

$$T_0 = T_E + Ce^{k(0)}$$
$$T_0 = T_E + Ce^0$$
$$T_0 = T_E + C(1)$$
$$T_0 - T_E = C$$

Substitute this value of C into your solution:

$$T = T_E + Ce^{kt}$$
$$T = T_E + (T_0 - T_E)e^{kt}$$

The final equation is the solution you're looking for. It translates Newton's Law of Cooling into a handy formula that you can use any time you want. For example, you could use it in part (b) of the problem!

KELLEY'S CAUTIONS

This problem is just riddled with *t*'s. There's *t, T, dt, dT, T₀, T_E*, and there's even *tea* in the cup! Proceed carefully or your answer might *dt*-iorate right before your eyes.

(b) Begin by taking inventory of the information you are given. You know that the original temperature of your tea was $T_0 = 140$ in a room with temperature $T_E = 72$. After $t = 17.5$ minutes, the tea has a new temperature of $t = 113$. Plug all of this information into the equation you created in part (a).

$$T = T_E + \left(T_0 - T_E\right)e^{kt}$$
$$113 = 72 + \left(140 - 72\right)e^{k(17.5)}$$
$$113 - 72 = \left(68\right)e^{17.5k}$$
$$41 = 68e^{17.5k}$$

The only information the problem does not give you is the unique value of k for this situation. Luckily, you can solve this equation to determine the value of k:

$$\frac{41}{68} = e^{17.5k}$$
$$\ln\left(\frac{41}{68}\right) = \ln\left(e^{17.5k}\right)$$
$$\ln\left(\frac{41}{68}\right) = 17.5k$$
$$\frac{\ln\left(41/68\right)}{17.5} = k$$
$$-0.028910608 \approx k$$

Now that you know the value of k, you can finally take the t out of "mystery." Come to think of it, that would leave you with "mysery," which sounds like "misery," which is an apt description for a problem this long.

You want to know what time t elapses between your drink's original temperature of $T_0 = 140$ and the new temperature $T = 125$. You already know that $T_E = 72$ and you just discovered a shiny new value of k to use. Let's break out Newton's Law of Cooling one last time:

$$T = T_E + \left(T_0 - T_E\right)e^{kt}$$
$$125 = 72 + \left(140 - 72\right)e^{(-0.028910608)t}$$
$$125 - 72 = 68e^{-0.028910608t}$$
$$53 = 68e^{-0.028910608t}$$
$$\frac{53}{68} = e^{-0.028910608t}$$
$$\ln\left(\tfrac{53}{68}\right) = -0.028910608t$$
$$\frac{\ln(53/68)}{-0.028910608} = t$$
$$8.62 \approx t$$

Your tea cools to 125°F after approximately 8.62 minutes. There are 60 seconds in a minute, so 0.62 minutes is equal to $(0.62)(60) = 37.2$ seconds. Therefore, your tea reaches 125°F after approximately 8 minutes, 37.2 seconds.

YOU'VE GOT PROBLEMS

Problem 4: Those big members-only warehouse superstores always sell things in such gigantic quantities. It's unclear what possessed you to buy 15,000 grams of Radon-222 radioactive waste. Perhaps you thought it would complement your 50-gallon barrel of mustard. In any case, it was a bigger mistake to drop it in the parking lot.

All radioactive waste has a defined half-life—the period of time it takes for half of the mass of the substance to decay away. The half-life of Radon-222 is 3.82 days. (In other words, 3.82 days after the waste pours out on the asphalt, 7,500 grams remain, and only 3,750 grams 3.82 days after that.) Approximately how long will it take for the 15,000 grams of Radon-222 to decay to a harmless 50 grams?

The Least You Need to Know

- Differential equations contain derivatives; solutions to basic differential equations are simply the antiderivatives solved for y.

- If a problem contains sufficient information, you can find a specific solution for a differential equation; it won't contain a "$+ C$" term.

- If a population's rate of growth or rate of decay is proportional to the size of the population, the growth or decay is exponential in nature.

- Exponential growth and decay are modeled with the equation $y = Ne^{kt}$.

Visualizing Differential Equations

Our brief encounter with differential equations is almost at an end. Like ships passing in the night, we will soon go our separate ways, and all you'll have left are the memories. Before you get too nostalgic, though, we have to discuss some slightly more complex differential equation topics.

We'll start with linear approximation, which we actually could have discussed in Chapter 9, because it is basically an in-depth look at tangent lines. However, it is a precursor to a more complex topic called Euler's Method, which is an arithmetic-heavy way to solve differential equations if you can't use separation of variables. It's a good approximation technique to have handy, because there are about 10 gijillion other ways to solve differential equations that we don't know the first thing about at this level of mathematical maturity.

Before we call it quits, we'll also spend some quality time with slope fields. They are exactly what they sound like—a field of teeny little slopes, planted like cabbages. By examining those cabbages, we can tell a lot about the solution to a differential equation. All in all, this chapter focuses on ways to broaden our understanding of differential equations without having to learn a whole lot more mathematics in the process. You have to love that.

In This Chapter

- Approximating function values with tangent lines
- Slope fields: functional fingerprints
- Using Euler's Method to solve differential equations
- Drawing slope fields with a calculator

Linear Approximation

We've been finding derivatives like mad throughout this entire book. Even though the derivative is the slope of the tangent line, it took us a while to appreciate why that could be even a remotely useful thing to know. In time, we learned that derivatives describe rates of change and can be used to optimize functions, among other things.

Let's add something new to the list about how mind-numbingly useful derivatives are. Take a look at the graph of a function $f(x)$ and its tangent line at the point $(c, f(c))$ in Figure 18.1.

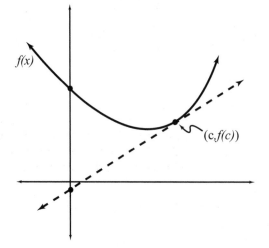

Figure 18.1

The graphs are very close to one another at $x = c$, but the father away from the point of tangency, the father apart they get. The line, for example, would not give you a good approximate value for $f(0)$. The y-intercept of the line is negative, but the function is positive when $x = 0$.

Notice how the graph of $f(x)$ and the tangent line graph get very close to each other around $x = c$. If you were to plug a value of x very close to c into both functions, you'd get almost the same output.

Because the equation of a tangent line to a function has values that usually come very close to the function around the point of tangency, that tangent equation is a good *linear approximation* for the function. No matter how simple a function may be to evaluate, not many functions are simpler than the equation of a line. Furthermore, some functions are way too hard to evaluate without a calculator, and linear approximations come in handy for approximating such functions.

 DEFINITION

A **linear approximation** is the equation of a tangent line to a function used to approximate the function's values lying close to the point of tangency.

Example 1: Estimate the value of ln(1.1) using the linear approximation to $f(x) = \ln x$ centered at $x = 1$.

Solution: The problem asks you to center your linear approximation at $x = 1$; this means that you should find the equation of the tangent line to $f(x)$ at that x-value. It's easy to build the equation of a tangent line—you did it way back in Chapter 10. All you need is the slope of the tangent line, which is $f'(1) = \frac{1}{1} = 1$, and the point of tangency, which is $f(1) = \ln 1 = 0$.

With this information, use the point-slope equation to build the tangent line:

$$y - y_1 = m(x - x_1)$$
$$y - 0 = 1(x - 1)$$
$$y = x - 1$$

KELLEY'S CAUTIONS

Remember: use a linear approximation only for x-values close to the x at which the approximation was centered. Notice that the approximation in Example 1 gives an awful approximation of $\ln x$ when $x = \frac{1}{8}$: -0.87. The actual value of $\ln \frac{1}{8}$ is -2.079. That's rather inaccurate, even though $\frac{1}{8}$ is less than one unit away from the center of the approximation!

Now plug $x = 1.1$ into the linear equation; you'll get $y = 1.1 - 1 = 0.1$. The actual value of ln(1.1) is 0.09531, so your estimate is pretty close.

YOU'VE GOT PROBLEMS

Problem 1: Estimate the value of arctan (1.9) using a linear approximation centered at $x = 2$.

Slope Fields

Even if you can't solve a differential equation, you can still get a good idea of what the solution's graph looks like. You just learned that a graph's tangent line looks a lot like the graph right around the point of tangency. Well, if you draw little tiny pieces of tangent line all over the coordinate plane, those pieces will show the shape of the solution graph. It's similar to using metallic shavings to determine where magnetic fields lie, or sprinkling fingerprint powder on a table's surface to highlight the shape of the print.

Drawing a *slope field* is a very simple process for basic differential equations, but it can get a bit repetitive. All you have to do is plug points from the coordinate plane into the differential equation. Remember, the differential equation represents the slope of the solution graph, as it is the first derivative. You will then draw a small line segment centered at that point with the slope you calculated.

DEFINITION

A **slope field** is a tool to help you visualize the solution to a differential equation. It is made up of a collection of line segments centered at points whose slopes are the values of the differential equation evaluated at those points.

Let's start with a very basic example: $\frac{dy}{dx} = 2x$. You know that the solution to this differential equation is $y = x^2 + C$, a family of parabolas with their vertices on the y-axis. Let's draw the slope field for $\frac{dy}{dx} = 2x$. First, let's identify the fertile field where our slopes will grow and flourish (see Figure 18.2).

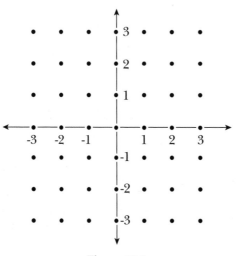

Figure 18.2

Each of the points indicated has coordinates that are integers; this makes the substitution a little quicker and easier.

At every dot on that field, you're going to draw a tiny little segment. Let's start at the origin. If you plug (0,0) into $\frac{dy}{dx} = 2x$, you get $\frac{dy}{dx} = 2 \cdot 0 = 0$, so the slope of the tangent line there will be 0 (i.e., the line is horizontal). Therefore, draw a small horizontal segment centered at the origin. The substitution was pretty easy—you didn't even have to plug the y-value in, because there is no y in the differential equation.

Now, you should do the same thing for every other point in Figure 18.2. Let's do one more together to make sure you've got the hang of this. For fun, I'll pick the point (1,2)—doesn't that *sound* fun? Plugging it into the differential equation gives you $\frac{dy}{dx} = 2 \cdot 1 = 2$. Therefore, the line segment centered at (1,2) will have a slope of 2.

Once you've plugged all those points into the differential equation, you should end up with something like Figure 18.3.

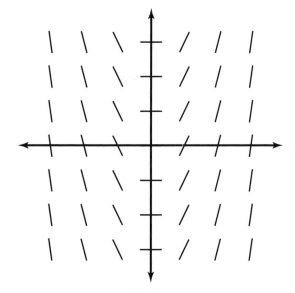

Figure 18.3
A slope field with just a hint of parabola.

 **CRITICAL POINT**

As a rule of thumb, a slope of 1 means a segment with a 45-degree angle. Greater slopes will be steeper and smaller slopes will be shallower. Negative slopes will fall from left to right, whereas positive slopes rise from left to right.

Can you see the shape hiding among all the little twigs? It's not perfect, but these little segments do a pretty good job of outlining the shape of a parabola whose vertex is on the y-axis. The slope field traces the shape of its solution curve. If you use a computer to draw the slope field, the shape is even clearer. (Computers don't tire as easily as I do when plugging in points; they don't even mind fractions.) For example, Figure 18.4 is a computer-generated slope field for $\frac{dy}{dx} = 2x$ with a specific solution shown, so you can see how well the slope field traces the solution.

 KELLEY'S CAUTIONS

A slope field will always outline a family of solutions. If you are given a point on the graph of the solution, place your pencil there and follow the paths of the slope segments to get an approximate graph of the specific solution.

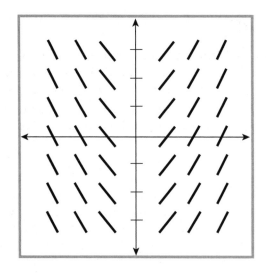

Figure 18.4

Although one parabola is drawn as a possible solution, it's easy to see that there are a lot of possible parabolas hiding in the woodwork. The solution graph assumes that the point (0,−1) is a point on the solution.

Again, this is a very detailed slope field; the computer calculates slopes at many fractional coordinates as well as integer coordinates. If you know that the solution to the differential equation contains the point (0,−1), you get the specific solution, represented by the darkened graph.

> ✏️ **CRITICAL POINT**
>
> You can download a fantastic program called *Graphmatica* that draws slope fields on your computer at graphmatica.com.

Slope fields are most useful when you can't solve the given differential equation by separation of variables but still want to see what the graph of the solution looks like. There are many differential equations out there that we can't solve at this level of our journey toward mathematical enlightenment, so it's good to enlist as many allies as we can.

> **YOU'VE GOT PROBLEMS**
>
> Problem 2: Draw the slope field for $\frac{dy}{dx} = \frac{x+y}{x-y}$ and sketch the specific solution to the differential equation that contains the point (0,1).

Euler's Method

To understand what *Euler's Method* really accomplishes in the land of differential equations, we need to talk about navigating through the woods. I'm not a huge fan of the outdoors, to be perfectly honest. I'm glad to be inside with the air conditioning on, away from flies, ticks, and those ugly little spiders that burrow under your skin and lay eggs in your brain. You may say those kinds of spiders don't exist and you're probably right, but there's nothing wrong with being too careful.

> **DEFINITION**
>
> **Euler's Method** is a technique used to approximate values on the solution graph to a differential equation when you can't actually find the specific solution to the differential equation via separation of variables. By the way, Euler is pronounced *OIL-er,* not *YOU-ler.*

This was not the case when I was younger. I always enjoyed tromping around outside and coming in as dirty as possible, covered in mud, sand, grass stains, and mashed bugs. In particular, I enjoyed exploring the woods with my friends. Most of the time, we'd be in areas either my friends or I knew extremely well. We even had crude maps of the woods drawn out, not that we ever actually had to resort to them. In new or unfamiliar woods, however, we'd rely on a compass to direct us to a road we knew: "Okay, if we get lost we'll go west and meet at the rotten tree trunk John lost his shoe in last year; watch out for those brain-egg spiders along the way."

When you solve a differential equation using separation of variables, you're given a map to all the correct solutions for that differential equation. In fact, the correct path to follow is the graph of the equation's solution. Back to our simple example from earlier: if the solution to the differential equation $\frac{dy}{dx} = 2x$ contains the point (3,6), you can find the exact solution using separation of variables. The antiderivative will be $y = x^2 + C$, and you can plug in the coordinate pair to find C:

$$6 = 3^2 + C$$
$$6 = 9 + C$$
$$-3 = C$$

Therefore, the exact solution to the differential equation is $y = x^2 - 3$. Now that you have this solution, it maps your way to other values on the solution graph. For example, it's very easy to determine what the value of $y(4)$ is (i.e., the solution graph's output when you input $x = 4$):

$$y = x^2 - 3$$
$$y(4) = 4^2 - 3$$
$$y(4) = 16 - 3$$
$$y(4) = 13$$

The solution graph "map" makes it easy to find the correct y-value corresponding to any x-value. But—and this is a big but—what if you can't solve the differential equation by separation of variables? Instead of a map, you'll use the compass of Euler's Method.

You'll still be given a point on the solution curve in these problems, but you won't be able to use it to find C. Instead, you'll use it as your reference point ("If you get lost, go west and meet at the shoe-devouring tree trunk"). From there, you'll take a compass reading ("We should go north—that big tree that looks like Scooby Doo is north from here"). When you have gone a fixed distance, you'll take another compass reading ("Okay, we're at the tree; now we should go northeast to that log with the frog on it"). After every small journey (as you reach each landmark), you'll take a compass reading and make sure that your course is true, and that you're heading in the right direction. After all, you are navigating in unknown lands without a map, with dangerous spiders everywhere. Better keep checking your compass.

Remember, a function and its tangent line have nearly equal values near the point of tangency. This is essential to Euler's Method. Taking compass readings in the woods is analogous to finding the correct derivative for the given function. We'll then step carefully along this slope for a fixed amount of time. If we go too far, the values of the slope will become too different from the values of the function (whose map we don't have). So we'll make another derivative check and start moving down this new direction. Remember, we don't know where the path is, but by using derivatives, we're staying as close to it as possible.

Before we can actually perform Euler's Method, we need to possess one prerequisite skill. Let's say you are at the point (0,3) and want to walk along a certain line that passes through that point. If that line has slope $m = \frac{2}{5}$, then walking up two units and to the right five units—arriving at the point (5,5)—ensures that you stay on the line. However, what if you only want to go $\frac{1}{3}$ of a unit up? How many units would you go right to make sure you were still on the line?

Example 2: Line l passes through (0,3) and has slope $m = \frac{2}{5}$. Without finding the equation of line l, find the correct y in the coordinate pair $\left(\frac{1}{3}, y\right)$ if that point is also on line l.

CRITICAL POINT

In Example 2, you're learning how to use a compass reading $\left(m=\frac{2}{5}\right)$ to walk a short distance $\left(x=\frac{1}{3}\right)$ and yet stay on the correct path.

Solution: The point $\left(\frac{1}{3},y\right)$ is exactly $\frac{1}{3}$ of a unit to the right of the original point $(0,3)$, so you can say that the change in x is $\frac{1}{3}$ from the first to the second point. Mathematically, this is written $\Delta x=\frac{1}{3}$. All you have to do is find the corresponding Δy to determine how far you should go vertically from the original y-value of 3. Remember that according to the slope equation you learned in your mathematical infancy, slope is equal to the change in y divided by the change in x: $m=\frac{y}{x}$. Use this equation to find the correct Δy:

$$m=\frac{y}{x}$$
$$\frac{2}{5}=\frac{y}{1/3}$$
$$5\ y=\frac{2}{3}$$
$$y=\frac{2}{3}\cdot\frac{1}{5}$$
$$y=\frac{2}{15}$$

According to this, you need to go up $\frac{2}{15}$ of a unit from the original y-value of 3 to stay on the line. Because $3+\frac{2}{15}=\frac{45}{15}+\frac{2}{15}=\frac{47}{15}$, the point $\left(\frac{1}{3},\frac{47}{15}\right)$ is guaranteed to be on the line through $(0,3)$ whose slope is $\frac{2}{5}$.

YOU'VE GOT PROBLEMS

Problem 3: If you begin at point $(-1,4)$ and proceed to the right a distance of $\Delta x=\frac{1}{2}$ along a line with a slope of $m=-\frac{2}{3}$, at what point do you arrive?

Now it's time to actually use Euler's Method. Euler's problems give you a differential equation, a starting point, and a value that needs estimating on the solution curve. You'll be told how many steps of what width to use, and you'll take steps of that width using the same method you used in Example 2.

Example 3: Use Euler's Method with three steps of width $\Delta x=\frac{1}{3}$ to approximate $y(3)$ if $\frac{dy}{dx}=x+y$ and the point $(2,1)$ appears on the solution graph.

Solution: It should be clear why the width of the steps is $\Delta x=\frac{1}{3}$—you're stepping from $x=2$ to $x=3$ in three steps. You'll repeat the same process three times, one for each step.

Step 1: From $x=2$ to $x=\frac{7}{3}$ (or $2\frac{1}{3}$)

The tangent slope at the point (2,1) is:

$$\frac{dy}{dx} = x + y$$
$$= 2 + 1$$
$$= 3$$

KELLEY'S CAUTIONS

The more steps you take (i.e., the smaller the width of each step), the more accurate your final approximation will be. Even with large steps, however, Euler's Method gets messy quickly; the ugly fractions compound, and it's easy to make an arithmetic mistake. It's best to check your work with a calculator.

Use this slope to calculate the correct value of Δy:

$$\frac{dy}{dx} = \frac{y}{x}$$
$$3 = \frac{y}{1/3}$$
$$y = 3 \cdot \frac{1}{3}$$
$$y = 1$$

This tells you to go up one unit from $y = 1$ while stepping right $\frac{1}{3}$ from $x = 2$:

$$\left(2 + \frac{1}{3}, 1 + 1\right) = \left(\frac{7}{3}, 2\right)$$

Step 2: From $x = \frac{7}{3}$ (or $2\frac{1}{3}$) to $x = \frac{8}{3}$ (or $2\frac{2}{3}$)

Repeat the same process as before, but use a starting point of $\left(\frac{7}{3}, 2\right)$ instead of (2,1). This time, the tangent slope is $\frac{dy}{dx} = x + y = \frac{7}{3} + 2 = \frac{13}{3}$ while Δx remains $\frac{1}{3}$. Find Δy:

$$\frac{13}{3} = \frac{y}{1/3}$$
$$3 \ y = \frac{13}{3}$$
$$y = \frac{13}{9}$$

So the starting point for the final step will be:

$$\left(\frac{7}{3} + \frac{1}{3}, 2 + \frac{13}{9}\right) = \left(\frac{8}{3}, \frac{31}{9}\right)$$

Step 3: From $x = \frac{8}{3}$ (or $2\frac{2}{3}$) to $x = 3$

This time, the slope of the tangent line is $\frac{dy}{dx} = \frac{8}{3} + \frac{31}{9} = \frac{55}{9}$; again, use it to find that $y = \frac{55}{27}$. Add Δx and Δy to $\left(\frac{8}{3}, \frac{31}{9}\right)$:

$$\left(\frac{8}{3} + \frac{1}{3}, \frac{31}{9} + \frac{55}{27}\right) = \left(3, \frac{148}{27}\right)$$

According to Euler's Method, the solution to the differential equation $\frac{dy}{dx} = x + y$ at $x = 3$ is approximately $\frac{148}{27}$, or 5.481.

YOU'VE GOT PROBLEMS

Problem 4: Use Euler's Method with three steps of width $x = \frac{1}{3}$ to approximate $y(1)$ if $\frac{dy}{dx} = 2x - y$ given that the solution graph passes through the origin.

Technology Focus: Slope Fields

I have good news and bad news. I'll start with the bad news: TI-84 calculators cannot generate slope fields natively. There is no function built into the factory software that allows you to draw them. However, there are plenty of programs you can download and install on your calculator if you are somewhat technologically savvy. Check out websites like ticalc.org for archives of programs including slope field generators.

Now for the better news: the TI-89 calculator *can* generate slope fields. The technique is a little counterintuitive, but it can be done. Let's start with Example 2 in this chapter, which asked you to draw the slope field for $\frac{dy}{dx} = 2x$ and then draw the specific solution to the differential equation that passed through point $(0,-1)$.

Start by pressing $y =$ and under the "Graph" option, change "Function" to "Diff Equations," as illustrated in Figure 18.5.

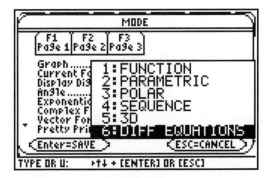

Figure 18.5

When you're finished graphing slope fields, be sure to change the mode back to "Function."

The familiar F1 screen, which you access by pressing "$\frac{dy}{dx} = 2x$," suddenly looks very unfamiliar. Because you are entering derivatives, the equations are labeled $y1'$ and $y2'$ instead of $Y1$ and $Y2$ (see Figure 18.6).

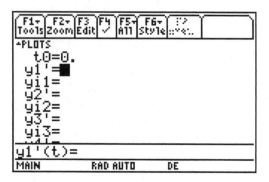

Figure 18.6
Enter your differential equations here.

You want to graph the slope field for $\frac{dy}{dx} = 2x$. Here's the strange part: use t instead of x when you type the equation. Because you also want the specific solution that equals −1 when $x = 0$, type −1 as the initial value for the differential equation: $yi1 = -1$ (see Figure 18.7).

```
F1▾  F2▾  F3  F4  F5▾  F6▾  ?▸
Tools Zoom Edit  ✓  All Style ::▾<:.

▸PLOTS
   t0=0.
 ✓y1'=2·t
   yi1=-1
   y2'=■
   yi2=
   y3'=
   yi3=

y2'(t)=
MAIN        RAD AUTO      DE
```

Figure 18.7
This will generate the slope field for $\frac{dy}{dx} = 2x$ that passes through point (0,−1).

Before you graph the slope field, be sure to adjust your graph window by pressing $\frac{dy}{dx} = \frac{x+y}{x-y}$ and selecting "ZoomSqr" or "Zoom Std." This ensures that the units are of equal length both horizontally and vertically. It keeps your graph in the correct proportions. The graph of the function (Figure 18.8) looks like the solution we expected, based on Figure 18.4.

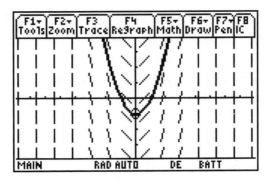

Figure 18.8
The slope field for $\frac{dy}{dx} = \frac{x+y}{x-y}$ and specific solution with initial (x = 0) value of −1.

How about something a little more complicated? If your differential equation contains x's and y's, the notation is, once again, sadly unintuitive. Let's use the calculator to check your solution to Problem 2 from the "You've Got Problems" sidebar earlier in this chapter. You're asked to draw the slope field for $\frac{dy}{dx} = \frac{x+y}{x-y}$ and then graph the specific solution that passes through (0,1).

When you type this into your calculator, once again replace x with t. In addition, you should type y as $y1$. You don't need to use any special characters—just type the y followed by a 1. You're also looking for the specific solution that equals 1 when $x = 0$, so enter a value of 1 for $yi1$ (see Figure 18.9).

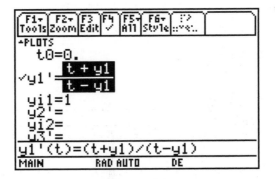

Figure 18.9
This will draw the slope field for $\frac{dy}{dx} = \frac{x+y}{x-y}$ and a specific solution that equals 1 when x = 0.

The resulting graph (Figure 18.10) matches the solution provided in Appendix A.

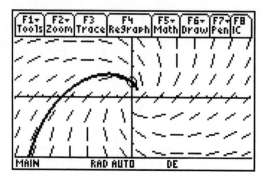

Figure 18.10

This graph is tricky to generate by hand, but the calculator can determine how best to draw the solution curve.

The Least You Need to Know

- A function and its tangent line have similar values near the point of tangency.
- Slope fields are collections of small tangent lines spread out over the coordinate plane that trace the graphs of solutions to differential equations.
- Euler's Method is used to approximate solutions to differential equations via linear approximation.
- Some calculators can draw slope fields, which is pretty handy for checking your work.

CHAPTER 19

Final Exam

Nothing helps you understand math like good old-fashioned practice, and that's the purpose of this chapter. You can use it however you like, but I suggest one of the following three strategies:

1. As you finish reading each chapter, skip back here and work on the practice problems from that chapter.

2. If you're using this book as a refresher for a class you've already taken, complete this test before you start reading the book. Then, go back and read the chapters containing problems you missed. After you've reviewed those topics, try these problems again.

3. Save this chapter until the end, and use it to see how much you remember of each topic when you haven't seen it for a while.

In This Chapter

- Measuring your understanding of all major calculus topics
- Practicing your skills
- Determining where you need more practice

Because these problems are just meant for practice, and not meant to teach new concepts, only the answers are given at the end of the chapter, usually without explanation or justification (unlike the problems in the "You've Got Problems" sidebars throughout the book). However, these practice problems are designed to mirror those examples, so you can always go back and review if you forgot something or need extra practice.

Are you ready? There's a lot of practice ahead of you—as some problems have multiple parts, there are actually more than 100 practice problems in this chapter! (But no one said you have to do them all at one sitting.)

Chapter 2

1. Put the linear equation in standard form: $-3(x + 2y) - 4y + 8 = x - 1$.

2. Determine the equation of the line that passes through the point $(-5,3)$ and has slope $-\frac{1}{2}$; write the equation in standard form.

3. Calculate the slope of the line that passes through points $(2,-3)$ and $(-5,-8)$.

4. Line n passes through the point $(2,-1)$ and is perpendicular to the line $3x - 5y = 2$. Write the equation of n in standard form.

5. Simplify the expression $\dfrac{\left(2xy^3\right)^3}{\left(9x^4y^2\right)^2}$.

6. Factor the expression completely: $32x^2 - 98y^2$.

7. Solve the equation $2x^2 - 16x = 22$ by completing the square.

8. Solve the equation by factoring: $x^2 - 256 = 0$.

9. Solve the equation using the quadratic formula: $3x^2 + 4x + 1$.

Chapter 3

10. If $f(x) = x^2 - 4x$, $g\left(x\right) = \sqrt{x}$, and $h(x) = x - 4$, evaluate $f(g(h(13)))$.

11. Determine what kind of symmetry, if any, is evident in the graph of $y = x^5 - x^3 + x - 5$.

12. Find the inverse function, $f^{-1}(x)$, if $f(x) = 5x - 3$; verify that $f(x)$ and $f^{-1}\left(x\right)$ are inverses by demonstrating that $f\left(f^{-1}\left(x\right)\right) = f^{-1}\left(f\left(x\right)\right)$.

13. Given the graph of function $j(x)$ and a table of values for the following function $k(x)$, calculate $j^{-1}(k(2))$. Assume that $j(x)$ has inverse function $j^{-1}(x)$; also assume that all points indicated on the graph of $j(x)$ have integer coordinates.

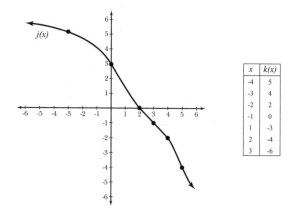

x	k(x)
-4	5
-3	4
-2	2
-1	0
1	-3
2	-4
3	-6

14. Put the parametric equations $x = 2t + 6$, $y = \sqrt{t - 1}$ into rectangular form.

Chapter 4

15. If $\cos\theta = -\dfrac{\sqrt{7}}{4}$ and $\sin\theta = -\dfrac{3}{4}$ calculate $\csc\theta$ and $\cot\theta$.

16. Evaluate $\sin\dfrac{14\pi}{3}$ using a coterminal angle and the unit circle.

17. Factor and simplify the trigonometric expression $1 - \tan^4\theta$.

18. Solve the equation $\left(2\cos x + \sqrt{2}\right)\left(\sin x - 1\right) = 0$ and provide all solutions on the interval $[0, 2\pi)$.

Chapter 5

19. Evaluate $\lim\limits_{x\to-8} \dfrac{x^2-64}{x+8}$.

20. Given $f(x)$ as defined here, calculate $\lim\limits_{x\to 4^+} f(x)$.

$$f(x)=\begin{cases} \sqrt{x}, & x\le 4 \\ x-5, & x>4 \end{cases}$$

21. Evaluate the limits on the graph:

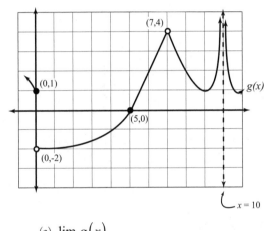

(a) $\lim\limits_{x\to 10} g(x)$

(b) $\lim\limits_{x\to 7} g(x)$

(c) $\lim\limits_{x\to 0} g(x)$

(d) $\lim\limits_{x\to 5} g(x)$

Chapter 6

22. Evaluate the limits using substitution:

(a) $\lim\limits_{x\to\pi/4} (x\cdot\sin x)$

(b) $\lim\limits_{x\to 2a} (x^2-3x+1)$

23. Evaluate the limits using the factoring method:

(a) $\lim\limits_{x\to 3/2} \dfrac{2x^2+7x+6}{2x+3}$

(b) $\lim\limits_{x\to-4} \dfrac{3x^2+11x-4}{5x^2+23x+12}$

24. Evaluate $\lim\limits_{x\to-4} \dfrac{\sqrt{x+5}-1}{x+4}$ using the conjugate method.

25. Evaluate the limit of $g(x)=\dfrac{x^2-11x+24}{x^2-2x-3}$ as x approaches each value for which $g(x)$ is undefined.

26. Evaluate the following limits:

(a) $\lim\limits_{x\to-\infty} \dfrac{9-x^2}{5+7x-3x^2}$

(b) $\lim\limits_{x\to\infty} \dfrac{4x^3+6x^2}{19x^2-5x+2}$

(c) $\lim\limits_{\theta\to 0} \dfrac{1-\cos 4\theta}{\theta}$

(d) $\lim\limits_{x\to 2} \left(\dfrac{x^2+4x-12}{x^2-11x+18} - \dfrac{3x}{x-4}\right)$

27. Given $h(x)=\tan x$, evaluate the following limits:

(a) $\lim\limits_{x\to 3\pi/4} h(x)$

(b) $\lim\limits_{x\to\pi/2} h(x)$

Chapter 7

28. Determine whether or not the function $f(x)$, as defined here, is continuous at $x=4$:

$$f(x)=\begin{cases} \dfrac{4x^2-13x-12}{x-4}, & x\ne 4 \\ 19, & x=4 \end{cases}$$

29. Find the value of c that makes the function $g(x)$ everywhere continuous:

$$g(x)=\begin{cases} x^2+3x-5, & x\le 1 \\ x-c, & x>1 \end{cases}$$

30. Find all the x-values for which the function $h(x)=\dfrac{x^2-12x+35}{2x^2-13x-7}$ is discontinuous and classify each instance of discontinuity.

31. Does the Intermediate Value Theorem guarantee the following function values for $f(x)=3x^2-12x+4$ on the closed interval $[0,5]$? Why or why not?

    (a) 10

    (b) 20

Chapter 8

32. Use the difference quotient to find the derivative of $f(x)=x^3-2x$ and use it to evaluate $f'(-3)$.

33. Determine $g'(1)$ if $g(x)=3x^2-8x+2$ using the alternative formula for the difference quotient.

Chapter 9

34. Find the derivative of each expression with respect to x:

    (a) $2x^5+6x^4-7x^3+\frac{1}{5}x^2-x+9$

    (b) $(3x^2+4)(9x-5)$

    (c) $\frac{2x^3+1}{x^2-4}$

    (d) $(x^2-7x+2)^{10}$

    (e) $\sqrt{x^3}\left(2x-3\right)^4$

35. Given the function $h(x)=3x^4-9x^2+2$, calculate the following values:

    (a) The average rate of change of $h(x)$ on the x-interval $[-1,3]$

    (b) The instantaneous rate of change of $h(x)$ when $x=2$

36. Given $f(x)=\tan(\cos x)$, calculate $f'\left(\frac{3\pi}{2}\right)$.

37. Assume functions $j(x)$ and $k(x)$ are continuous and differentiable for all real numbers. The following table lists values of the functions and their derivatives for specific x-values.

x	$j(x)$	$k(x)$	$j'(x)$	$k'(x)$
0	−1	3	1	−5
1	2	−1	3	−2
2	5	2	6	1
3	9	4	10	3
4	15	0	12	−3

Given this information, calculate the following:

    (a) $v'(3)$, given $v(x)=j\big(k(x)\big)$

    (b) $p'(0)$, given $p(x)=\dfrac{k(x)}{j(x)}$

Chapter 10

38. Identify the equation of the tangent line to $f(x)=x^2\sin x$ when $x=\pi$. *Hint:* use the Product Rule to differentiate $f(x)$.

39. Calculate the slope of the tangent line to the graph of $x^2-7xy-4y^2+y-9=0$ at the point $(-3,0)$.

40. Calculate the slope of the normal line to the graph of $j(x)=\tan x\cos x$ when $x=0$.

41. Given $g(x)=x^3-4$, evaluate $(g^{-1})'(-3)$.

42. If $h(x) = -2x^3 - 5x + 3$, calculate $(h^{-1})'(-1)$.

43. Given the parametric equations $x = \cos\theta$ and $y = 2\theta$, determine $\frac{dy}{dx}$ and $\frac{d^2y}{dx^2}$.

Chapter 11

44. If $f(x) = x^3 - 16x$, find $f'(x)$, determine its critical numbers, and determine if $f(x)$ changes direction at each.

45. If some function $g(x)$ has derivative $g'(x) = \frac{(x+3)(x-5)(x+4)}{x-1}$, use a wiggle graph to determine the interval(s) on which $g(x)$ is decreasing.

46. What are the absolute maximum and minimum values of $h(x) = \frac{x}{x-4}$ on the closed interval $[-4,3]$?

47. On what interval is $f(x) = x^3 - 8x^2 + 9x - 12$ concave up?

48. Use the Second Derivative Test to classify the relative extrema of the function $g(x) = x^3 + \frac{13}{2}x^2 - 30x + 19$.

Chapter 12

49. A goldfish swims back and forth inside a large fish tank featuring a plastic, bubbling, sunken treasure chest ornament. At time t, the horizontal position of the goldfish (relative to the treasure chest) is $s(t) = \frac{1}{9}t^2 - \cos 4t - 3$ inches. (If $s(t) > 0$, the fish is right of the treasure chest, and a negative $s(t)$ means the fish is left of it.)

Based on this information, answer the following questions:

(a) At what time(s) is the fish 3.5 inches left of the treasure chest?

(b) What is the speed of the fish at $t = 4.2$ seconds?

(c) What is the fish's average velocity between $t = 0$ and $t = 5$?

(d) On what interval(s) does the fish have positive acceleration between $t = 0$ and $t = 2$ seconds?

50. Nick throws a baseball straight up from an initial height of 3 feet, with a velocity of 25 ft/sec. Given this information, answer the following questions:

(a) What is the velocity of the ball $t = 1$ second after Nick throws it?

(b) When does the ball reach its maximum height?

(c) What is the maximum height of the ball?

(d) Assuming no one catches the throw, how long does the ball remain in the air before it hits the ground?

Chapter 13

51. Given $x_0 = 0$, apply Newton's Method to calculate x_1 and approximate the root of $f(x) = e^{3x} - 2$.

52. Evaluate $\lim\limits_{x \to \pi/2} \frac{\cos x}{2x - \pi}$.

53. Given the function $f(x) = 6x^2 - 2x + 3$, find the x-value that satisfies the Mean Value Theorem on the interval $[-1,1]$.

54. Erin and Sara are coworkers and exit their office at precisely the same time, 5 P.M. Erin walks due south from the office at a constant speed of 3.5 miles per hour. Sara bikes due west at a constant speed of 12 miles per hour. At what rate is the distance between Erin and Sara increasing at exactly 5:15 P.M.?

55. A farmer owns a plot of land whose western boundary is a river. He wishes to design a rectangular pasture but will only use fence for three of its sides, trusting the river to define the remaining side of the pasture, as illustrated here.

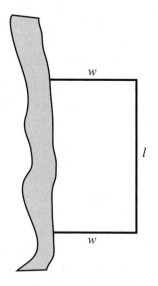

What are the dimensions of the largest pasture he can create using 2,500 feet of fence?

Chapter 14

56. Approximate the area under the curve $f(x) = \sqrt{x-3}$ on the interval $[4,8]$ using the following:

  (a) Right sums with $n = 8$ rectangles

  (b) Midpoint sums with $n = 4$ rectangles

  (c) Trapezoidal Rule with $n = 4$ trapezoids

  (d) Simpson's Rule with $n = 6$ subintervals

Chapter 15

57. Evaluate: $\int \left(x^5 - 4x^3 + \frac{x}{8} - 7 + 5\sqrt[3]{x^2} \right) dx$.

58. Integrate: $\int \frac{3x^2 - 4x + 1}{x} dx$.

59. Calculate the area beneath the curve $f(x) = \sqrt{x}$ on the interval $[4,8]$ using a definite interval.

60. Calculate the derivative: $\frac{d}{dy} \left(\int_{-4}^{y^3} \cos w \, dw \right)$.

61. Integrate using u-substitution:

  (a) $\int \sec 5x \, dx$

  (b) $\int_0^3 x \cdot e^{x^2} dx$

  (c) $\int \frac{x-6}{x+5} dx$

Chapter 16

62. Calculate the area between the functions $f(x) = \sqrt{x}$ and $g(x) = \frac{x}{3}$.

63. Calculate the value guaranteed by the Mean Value Theorem for Integration on the function $h(x) = \frac{1}{1+x^2}$ on the interval $[0,1]$.

64. The velocity of a particle moving horizontally along the x-axis is modeled by the equation $v(t) = t^3 - 7t + 6$, measured in inches per second. Use this information to answer the following questions:

    (a) What is the total displacement of the particle between $t = 0$ and $t = 3$?

    (b) What total distance does the particle travel between $t = 0$ and $t = 3$?

65. Given $f(x) = \int_0^{x^2} e^{3t}\, dt$, evaluate the following:

    (a) $f(1)$

    (b) $f'(1)$

66. Write an integral expression representing the length of each graph segment described here, and then use a computational tool (such as a graphing calculator) to compute each integral:

    (a) $f(x) = \tan x$, between $x = \frac{\pi}{6}$ and $x = \frac{\pi}{3}$

    (b) The parametric curve defined by $x = e2t$ and $y = \ln(4t + 2)$ on the t-interval $[0,3]$

Chapter 17

67. Solve the differential equation $x^2 dy = -2dx$.

68. A popular new song is predicted to sell at a rate of $\frac{dy}{dt} = \frac{10}{t^2+25}$ million purchases per day. In fact, it sells 1.85 million copies by the end of the first day alone! Use this information to answer the following questions:

    (a) What equation, $y(t)$, models the sales of this CD? *Note:* calculate C accurate to four decimal places.

    (b) Approximately how many songs will have been sold exactly 730 days (2 years) after release? *Note:* round your answer to four decimal places.

69. By ignoring any standards of cleanliness, and choosing to live a life of squalor, you have inadvertently invented a new kind of chemical weapon forged out of soggy Cheetos, stagnant milk-filled cereal bowls, and a chocolate Easter bunny of indeterminate age.

    Here's the downside. The government has quarantined you inside your filthy house until the nasty mixture disintegrates a bit. In a truly disturbing development, they've determined that (like nuclear waste), your food weapon has a half-life, and they're reasonably sure the half-life is four days.

    Assuming this is true, how long will it take the 3,000 grams of dangerous procrastination-produced glop to decay to a safer (but equally stinky) 10 grams?

Chapter 18

70. Estimate the value of $\sqrt{9.1}$ without a calculator by using a linear approximation to $f(x) = \sqrt{x}$ centered at $x = 9$.

71. Draw the slope field for $\frac{dy}{dx} = \frac{x}{y+1}$ by calculating slopes at each point indicated in the following coordinate plane:

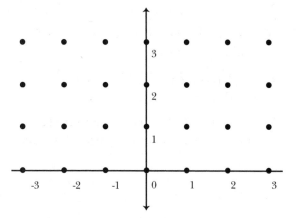

Sketch the specific solution to the differential equation that contains the point $(0,1)$.

72. If you begin at the point $\left(\frac{1}{3}, -2\right)$ and proceed $x = \frac{1}{5}$ units to the right along a line with slope $m = \frac{3}{8}$, what are the coordinates of your destination?

73. Use Euler's Method with three steps of width $x = \frac{1}{4}$ to approximate $y\left(-\frac{1}{4}\right)$ if $\frac{dy}{dx} = xy$, given that the solution graph passes through $(-1,1)$.

Solutions

Chapter 2

1. $4x + 10y = 9$

2. $x + 2y = 1$

3. $\frac{5}{7}$

4. $5x + 3y = 7$

5. $\frac{8y^5}{81x^5}$

6. $2(4x + 7y)(4x - 7y)$

7. $x = 4 - 3\sqrt{3}$ or $x = 4 + 3\sqrt{3}$

8. $x = -16$ or $x = 16$

9. $x = -1$ or $x = -\frac{1}{3}$

Chapter 3

10. -3

11. no symmetry

12. $f^{-1}(x) = \frac{1}{5}x + \frac{3}{5}$,

$\frac{1}{5}(5x - 3) + \frac{3}{5} = 5\left(\frac{1}{5}x + \frac{3}{5}\right) - 3 = x$

13. 5

14. $y = \sqrt{\frac{1}{2}x - 4}$

Chapter 4

15. $\csc\theta = -\frac{4}{3}$ and $\cot\theta = \frac{\sqrt{7}}{3}$

16. $\sin\frac{14\pi}{3} = \sin\frac{2\pi}{3} = \frac{\sqrt{3}}{2}$

17. $\sec^2\theta(1 + \tan\theta)(1 - \tan\theta)$

18. $x = \frac{\pi}{2}, \frac{3\pi}{4}, \frac{5\pi}{4}$

Chapter 5

19. −16

20. −1

21a. does not exist because
$\lim\limits_{x \to 10^+} g(x) = \lim\limits_{x \to 10^-} g(x) = \infty$, but ∞ is not a finite number

21b. 4

21c. does not exist because
$\left(\lim\limits_{x \to 0^-} g(x) \neq \lim\limits_{x \to 0^+} g(x) \right)$

21d. 0

Chapter 6

22a. $\frac{\pi\sqrt{2}}{8}$

22b. $4a^2 - 6a + 1$

23a. $\frac{7}{2}$

23b. $\frac{13}{17}$

24. $\frac{1}{2}$

25. $\lim\limits_{x \to 3} g(x) = -\frac{5}{4}$, $\lim\limits_{x \to -1} g(x)$ does not exist

26a. $\frac{1}{3}$

26b. $\lim\limits_{x \to \infty} \frac{4x^3 + 6x^2}{19x^2 - 5x + 2} = \infty$, which means it does not exist

26c. 0

26d. $\frac{13}{7}$: split the two terms into separate limits, apply factoring and substitution methods, and add the results

27a. −1, because $\tan\left(\frac{3\pi}{4}\right) = \frac{\sin(3\pi/4)}{\cos(3\pi/4)} = \frac{\sqrt{2}/2}{-\sqrt{2}/2} = -1$

27b. does not exist, because the graph of $y = \tan x$ has a vertical asymptote at $x = \pi/2$ (see Figure 4.4)

Chapter 7

28. Because $f(4) = \lim\limits_{x \to 4} f(x) = 19$, $f(x)$ is continuous at $x = 4$

29. $c = 2$

30. $x = -\frac{1}{2}$ (infinite discontinuity) and $x = 7$ (point discontinuity)

31a. yes, because $f(0) = 4$, $f(5) = 19$, and $4 \leq 10 \leq 19$

31b. no, because 20 does not fall between the function values of the endpoints $f(0) = 4$ and $f(5) = 19$

Chapter 8

32. $f'(x) = 3x^2 - 2$, $f'(-3) = 25$

33. $g'(1) = -2$

Chapter 9

34a. $10x^4 + 24x^3 - 21x^2 + \frac{2}{5}x - 1$

34b. $81x^2 - 30x + 36$

34c. $\frac{2x^4 - 24x^2 - 2x}{x^4 - 8x^2 + 16}$

34d. $10(x^2 - 7x + 2)^9 (2x - 7)$

34e. $\frac{3\sqrt{x}(2x-3)^4}{2} + 8x\sqrt{x}(2x - 3)^3$. *Note:* use the Product Rule and take the derivative of $(2x - 3)^4$ with the Chain Rule

35a. 42

35b. 60

36.
$$f'\left(\frac{3\pi}{2}\right) = \sec^2\left(\cos\frac{3\pi}{2}\right) \cdot \left(-\sin\frac{3\pi}{2}\right)$$
$$= \sec^2(0) \cdot -(-1)$$
$$= \frac{1}{\cos^2 0} \cdot 1$$
$$= \frac{1}{1} \cdot 1$$
$$= 1$$

37a. Apply the Chain Rule:

$$v'(3) = j'(k(3)) \cdot k'(3)$$
$$= j'(4) \cdot k'(3)$$
$$= 12 \cdot 3$$
$$= 36$$

37b. Apply the Quotient Rule:

$$p'(0) = \frac{j(0) \cdot k'(0) - k(0) \cdot j'(0)}{\left[j(0) \right]^2}$$
$$= \frac{(-1)(-5) - (3)(1)}{(-1)^2}$$
$$= \frac{5 - 3}{1}$$
$$= 2$$

Chapter 10

38. $y = -\pi^{2x + \pi 3}$. *Note:* $f(\pi) = 0$ and $f'(\pi) = -\pi^2$

39. $y' = \frac{-2x + 7y}{-7x - 8y + 1} = \frac{6}{22} = \frac{3}{11}$

40. $j'(x) = -\tan x \sin x + \cos x \sec 2\, x$, so $j'(0) = 0 + 1 = 1$, which means tangent and normal slopes are both 1 (as 1 is the reciprocal of itself)

41. $\frac{1}{3}$

42. $\left(h^{-1} \right)'(1) = \frac{1}{-6(0.676280053)^2 - 5} \approx -0.129$

43. $\frac{dy}{dx} = \frac{2}{-\sin\theta} = -2\csc\theta$,
$\frac{d^2 y}{dx^2} = -\frac{2\cos\theta}{\sin^3\theta} = -2\csc^2\theta\cot\theta$

Chapter 11

44. $f(x)$ changes from increasing to decreasing at $x = -\frac{4}{\sqrt{3}} \approx -2.309$, because $f'(x)$ changes from positive to negative there; similarly, $f(x)$ changes from decreasing to increasing at $x = \frac{4}{\sqrt{3}} \approx 2.309$, because $f'(x)$ changes from negative to positive there

45. $g(x)$ decreases on $(-4, -3)$ and $(1, 5)$

46. maximum $= 0$, minimum $= -9$

47. $\left(\frac{8}{3}, \infty \right)$

48. $x = \frac{5}{3}$ is a relative minimum because $g''\left(\frac{5}{3} \right) = 23 > 0$, $x = -6$ is a relative maximum because $g''(-6) = -23 < 0$.

Chapter 12

49a. $t = 0.2596$, $t = 1.3756$, and $t = 1.7194$ seconds

49b. $\left| s'(4.2) \right| = \left| \frac{2}{9}(4.2) + 4\sin(16.8) \right| \approx \left| -2.617 \right| = 2.617$ in/sec

49c. 0.6739 in/sec

49d. $(0, 0.3962)$ and $(1.1746, 1.9670)$

50a. $v(t) = s'(t) = -32t + 25$, so $v(1) = -32(1) + 25 = -7$ ft/sec (the ball is falling at a rate of 7 ft/sec, so this tells you the ball has already reached its maximum height before $t = 1$)

50b. $v(t) = 0$ when $-32t + 25 = 0$, or $t = \frac{25}{32} = 0.78125$

50c. maximum height: $s(0.78125) \approx 12.7656$ feet

50d. Solve equation $-16t^2 + 25t + 3 = 0$ for t and select only the positive solution: $t \approx 1.674$ seconds.

Chapter 13

51. $x_1 = 0 - \frac{e^0 - 2}{3e^0} = 0 - \frac{1 - 2}{3 \cdot 1} = -\frac{-1}{3} = \frac{1}{3}$

52. $-\frac{1}{2}$

53. $x = 0$

54. After 15 minutes, Erin has traveled $\frac{3.5}{4} = 0.875$ miles and Sara has traveled $\frac{12}{4} = 3$ miles, so allow those to be the lengths of two sides of the right triangle illustrated here:

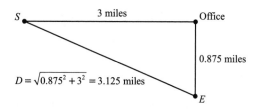

$D = \sqrt{0.875^2 + 3^2} = 3.125$ miles

Apply the Pythagorean Theorem to determine that the length D of the hypotenuse, the distance between them, is 3.125 miles and then differentiate the Pythagorean Theorem with respect to t:

$$E^2 + S^2 = D^2$$
$$2E\frac{dE}{dt} + 2S\frac{dS}{dt} = 2D\frac{dD}{dt}$$
$$2(0.875)(3.5) + 2(3)(12) = 2(3.125)\frac{dD}{dt}$$

Solve for $\frac{dD}{dt}$ to get 12.5 miles per hour.

55. $w = 625$ feet, $l = 2500 - 2(625) = 1250$ feet. *Note:* pasture perimeter is $2w + l = 2500$ so $l = 2500 - 2w$, plug this into the primary equation $A = lw$ to get $A = (2500 - 2w)w$ and optimize

Chapter 14

56a. $\frac{1}{2}\left[f\left(\frac{9}{2}\right) + f(5) + f\left(\frac{11}{2}\right) + f(6) + f\left(\frac{13}{2}\right) + f(7) + f\left(\frac{15}{2}\right) + f(8)\right] \approx 7.090$

56b. $1\left[f\left(\frac{9}{2}\right) + f\left(\frac{11}{2}\right) + f\left(\frac{13}{2}\right) + f\left(\frac{15}{2}\right)\right] \approx 6.798$

56c. $\frac{1}{2}\left[1 + 2\sqrt{2} + 2\sqrt{3} + 4 + \sqrt{5}\right] \approx 6.764$

56d. $\frac{8-4}{3(6)}\left[f(4) + 4f\left(\frac{14}{3}\right) + 2f\left(\frac{16}{3}\right) + 4f(6) + 2f\left(\frac{20}{3}\right) + 4f\left(\frac{22}{3}\right) + f(8)\right] \approx 6.787$

Chapter 15

57. $\frac{x^6}{6} - x^4 + \frac{1}{16}x^2 - 7x + 3x^{5/3} + C$

58. $\frac{3}{2}x^2 - 4x + \ln|x| + C$

59. $\frac{32\sqrt{2}}{3} - \frac{16}{3}$

60. $3y^2 \cos y^3$

61a. $\frac{1}{5}\ln|\sec 5x + \tan 5x| + C$. *Note:* set $u = 5x$

61b. $\frac{1}{2}\left(e^9 - 1\right)$

61c. $x - 11\ln|x + 5| + C$

Chapter 16

62. $\int_0^9 \left(x^{1/2} - \frac{1}{3}x\right) dx = \frac{9}{2}$

63. $h(c) = \frac{\int_0^1 \frac{dx}{1+x^2}}{1-0} = \arctan x\Big|_0^1 = \frac{\pi}{4}$

64a. $\int_0^3 v(t)\, dt = \frac{27}{4} = 6.75$

64b. $\int_0^1 v(t)\, dt - \int_1^2 v(t)\, dt + \int_2^3 v(t)\, dt = \frac{33}{4} = 8.25$

65a. $f(1) = \int_0^{12} e^{3t}\, dt = \frac{e^3}{3} - \frac{1}{3}$

65b. $f'(1) = e^{3\left(x^2\right)} \cdot 2x = 2e^3$

66a. $\int_{\pi/6}^{\pi/3} \sqrt{1 + \sec^4 x}\, dx \approx 1.277$. *Note:* you may need to enter $\sec^4 x$ as $(1/\cos(x))^4$

66b. $\int_0^3 \sqrt{\left(2e^{2t}\right)^2 + \left(\frac{4}{4t+2}\right)^2}\, dt \approx 402.616$

Chapter 17

67. $y = \frac{2}{x} + C$

68a. $y(t) = 2\arctan\left(\frac{t}{5}\right) + 1.4552$. *Note:*
$y(t) = \int \frac{10\, dt}{t^2 + 25} = 10 \cdot \frac{1}{5} \cdot \arctan\left(\frac{t}{5}\right) + C$ and $y(1) = 1.85$, so you need to solve the equation $1.85 = 2\arctan\left(\frac{1}{5}\right) + C$ for C

68b. $y(730) \approx 4.5831$, so 4,583,100 copies sold in two years

69. 32.915 days. *Note:* $y(t) = 3000e^{-0.173287t}$

Chapter 18

70. 3.01667. *Note:* linear approximation is $y = \frac{1}{6}x + \frac{3}{2}$

71.

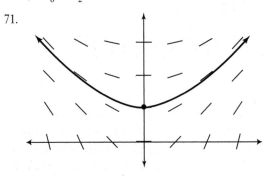

72. $\left(\frac{1}{3} + \frac{1}{5}, -2 + \frac{3}{40}\right) = \left(\frac{8}{15}, -\frac{77}{40}\right)$

73. $y\left(-\frac{1}{4}\right) = \frac{273}{512}$. *Note:* the coordinates of the three steps are $\left(-1 + \frac{1}{4}, 1 - \frac{1}{4}\right) = \left(-\frac{3}{4}, \frac{3}{4}\right)$, $\left(-\frac{3}{4} + \frac{1}{4}, \frac{3}{4} - \frac{9}{64}\right) = \left(-\frac{1}{2}, \frac{39}{64}\right)$, and $\left(-\frac{1}{2} + \frac{1}{4}, \frac{39}{64} - \frac{39}{512}\right) = \left(-\frac{1}{4}, \frac{273}{512}\right)$

Solutions to "You've Got Problems"

All of the answers to the problems that haunted you throughout the book are listed here, organized by chapter. The important steps are all shown, unless the skill needed to complete a problem was already discussed in a previous chapter. For example, once you learn how to do *u*-substitution in Chapter 15, I no longer focus on its details if problems in subsequent chapters require *u*-substitution as a component of their answers. If I didn't do that, this appendix would be a book unto itself!

Chapter 2

1. $6x + 9y = 11$. Don't forget that $6x$ has to be positive to be in standard form. You may need to multiply everything by -1.

2. $2x - 3y = 6$. You can treat $(0,-2)$ as a point or use it as the *y*-intercept, so both forms work.

3. $\frac{3}{4}$. Remember that $\frac{-3}{-4} = \frac{3}{4}$.

4. $\frac{11}{5}$. Calculate the slope of line *j*:
    $m = \frac{3-1}{4-(-6)} = \frac{2}{10} = \frac{1}{5}$. Substitute the slope and a point, such as $(x,y) = (4,3)$, into slope-intercept form ($y = mx + b$): $3 = \frac{1}{5}(4) + b$. Solve: $3 - \frac{4}{5} = b = 2\frac{1}{5}$ or $\frac{11}{5}$. Remember that *b* represents the *y*-intercept in slope-intercept form.

5. $\frac{9y^4}{x^6}$. When you square everything, you get $9x^{6}y^4$, and the term with the negative exponent has to be moved to the denominator.

6. $7xy(x - 3y^2)$. The greatest common factor is $7xy$, so divide it out of each term to get the factored form and write $7xy$ in front.

7. $(2x + 7)(4x^2 - 14x + 49)$. This is a sum of perfect cubes, with $a = 2x$ and $b = 7$.

8. $x = 0,-4$. **Method one:** Factor out $3x$. **Method two:** First divide by 3 to get $x^2 + 4x = 0$. Half of 4 is 2, whose square, 4, should be added to both sides. Take the square root of both sides, and don't forget the "$\pm$" symbol. Then subtract 2 from both sides. **Method three:** $a = 3$, $b = 12$, and $c = 0$, because there is no constant term.

9. $y = 3x^2 + x - 2$. Set the equations $x = -1$ and $x = \frac{2}{3}$ equal to 0 to get $x + 1 = 0$ and $x - \frac{2}{3} = 0$. To eliminate fractions, multiply the second equation by 3 to get $3x - 2 = 0$. The quadratic equation you are seeking is $y = (x + 1)(3x - 2)$. Multiply and simplify.

Chapter 3

1. 4. $f(43) = 7$; $g(7) = 64$; $h(64) = 4$.

2. $g(-2) = 8$, $g(0) = 12$, $g(5) = 4$. Because $x = -2$ and $x = 0$ are both less than or equal to 0, substitute them into $12 - x^2$. Substitute 5 into $\sqrt{x^2 - 9}$.

3. 9. The graph of $q(x)$ passes through point $(1,-5)$, so $q(1) = -5$. Next, $p(-5) = 4 - (-5) = 4 + 5 = 9$.

4. Origin symmetry. Plug in $-x$ for x and $-y$ for y to get $-y = \frac{-x^3}{|x|}$. Multiply both sides by -1, and you'll get the original function.

5. $\frac{1}{2}\left(\sqrt{2x+6}\right)^2 - 3 = \sqrt{2\left(\frac{1}{2}x^2 - 3\right) + 6} = x$

   $\frac{1}{2}(2x+6) - 3 \qquad = \sqrt{x^2 - 6 + 6} \qquad = x$

   $x + 3 - 3 \qquad\qquad = \sqrt{x^2} \qquad\qquad = x$

   $x \qquad\qquad\qquad = x \qquad\qquad\quad = x$

6. $h^{-1}(x) = \frac{3}{2}x - \frac{15}{2}$. After switching x and y, subtract 5 from both sides and then eliminate $\frac{2}{3}$ by multiplying each side of the equation by $\frac{3}{2}$.

7. $y = x^2 - 3x + 3$. Start by solving the x equation for t ($t = x - 1$), plug that into both t spots in the y equation, and simplify.

Chapter 4

1. 0. Simplify $\frac{14\pi}{4}$ to get $\frac{7\pi}{2}$. Subtract 2π (or $\frac{4\pi}{2}$) to get $\frac{3\pi}{2}$; $\cos\frac{3\pi}{2} = \cos\frac{14\pi}{4} = 0$.

2. $\left(-\frac{1}{2}\right)^2 + \left(\frac{\sqrt{3}}{2}\right)^2 = \frac{1}{4} + \frac{3}{4} = 1$

3. $\sin 2x \cos 2x$. Factor out the greatest common factor of $2\sin x \cos x$ to get $2\sin x \cos x (1 - 2\sin^2 x)$ and use double angle formulas to substitute in replacements for each factor.

4. $x = 0,\pi$. Substitute $2\sin x \cos x$ for $\sin 2x$ and factor to get $2\sin x (\cos x + 1) = 0$; solve each equation set equal to 0.

Chapter 5

8. Factor the numerator: $j(x) = \frac{(x+5)(x-3)}{x-3}$. Thus, $j(x) = x + 5$ is equivalent for all x-values except $x = 3$, as illustrated in the following graph.

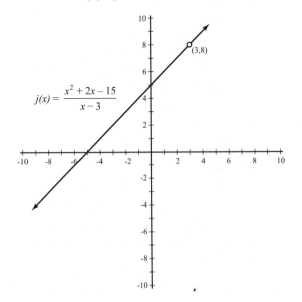

Substitute $x = 3$ into the simplified version of $j(x)$ to compute $\lim\limits_{x \to 3} j(x) : j(3) = 3 + 5 = 8$.

2. (a) $\lim\limits_{x \to 1^-} k(x) = 4$. As you approach $x = 1$ from the left, $k(x)$ intends to reach point $(1,4)$.

(b) $\lim\limits_{x \to 1^+} k(x) = -4$. As you approach $x = 1$ from the right, $k(x)$ intends to reach point $(1,-4)$.

3. (a) $-\infty$. The graph decreases infinitely as you approach $x = -4$ from the left. You can also answer that no limit exists because the graph decreases infinitely—both methods of answering are equivalent.

(b) Does not exist. The left-hand limit (-2) does not equal the right-hand limit (3), so no general limit exists.

(c) 1. The left- and right-hand limits are both 1, so the general limit exists and is 1.

Chapter 6

1. (a) $-\frac{1}{\pi}$. Plug in π for each x to get $\frac{\cos \pi}{\pi}$; you know that $\cos \pi = -1$ from the unit circle.

(b) $\lim\limits_{x \to -2} \frac{x^2+1}{x^2-1} = \frac{(-2)^2+1}{(-2)^2-1} = \frac{4+1}{4-1} = \frac{5}{3}$.

2. (a) 13. Factor the numerator to get $(2x + 3)$ $(x - 5)$; cancel the $(x - 5)$ terms and plug $x = 5$ into $2x + 3$.

(b) 3. The numerator is the difference of perfect cubes (remember the formula?), which factors to $(x - 1)(x^2 + x + 1)$; the $(x - 1)$ terms cancel, leaving only $x^2 + x + 1$; substitute $x = 1$ into that expression to get the answer.

3. (a) 4. Multiply numerator and denominator by $\sqrt{x+6}+2$ and cancel out resulting $(x + 2)$ terms to get $\lim\limits_{x \to -2} \left(\sqrt{x+6}+2\right)$; substitute $x = -2$ into the expression: $\sqrt{-2+6}+2 = \left(\sqrt{4}+2\right) = \left(2+2\right)$.

(b) $\frac{\sqrt{5}-3}{2}$. Did I fool you? You don't use the conjugate method here, because substitution works; to get the answer, just plug in $x = 1$ for all x's (no simplifying can be done).

4. Factor to get $\frac{x(2x-1)(x-1)}{x(2x-1)(x+3)}$. The function is undefined at $x = 0$, $x = \frac{1}{2}$, and $x = -3$. Using the factoring method, $\lim\limits_{x \to 0} g(x) = -\frac{1}{3}$ and $\lim\limits_{x \to 1/2} g(x) = -\frac{1}{7}$, so holes exist on the graph for those values. However, no limit exists for $x = -3$, because substitution results in $-\frac{84}{0}$, indicating that $x = -3$ is a vertical asymptote.

5. (a) $\frac{2}{3}$. The degrees of the numerator and denominator are the same.

(b) 0. The denominator has the higher degree; the fact that you're approaching $-\infty$ doesn't matter, because all rational functions possessing an infinite limit approach the same height as x approaches ∞ and $-\infty$.

6. $h(x) = \frac{1}{x-3}$ is one possible solution, but answers will vary. Here's how to check a different answer. First, $\lim\limits_{x \to \infty} h(x) = 0$, so $h(x)$ has a horizontal asymptote of $y = 0$. That means the degree of the numerator of $h(x)$ must be lower than the degree of the denominator. In my solution, the degree of the numerator is 0 (because 1 is a constant) and the degree of the denominator is 1 (because $x - 3$ is linear).

Second, $\lim_{x\to 3} h(x)$ can't exist. In my function, that limit does not exist because $h(x)$ has a vertical asymptote at $x = 3$. When you substitute $x = 3$ into $h(x)$, you get a 0 in the denominator but not in the numerator—an indication of a vertical asymptote.

You should also verify your solution with a graph. The graph of my solution, $h(x) = \frac{1}{x-3}$, appears here:

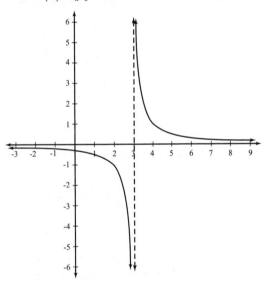

7. *e.* Break into two limits to get $\lim_{x\to\infty} \frac{5}{x^3} + \lim_{x\to\infty}\left(1+\frac{1}{x}\right)^x$; each of these is a separate special limit rule. The first limit is equal to 0 (by the third rule) and the other limit is equal to e (by the last rule), so the answer is $0 + e = e$.

8. (a) 0. The graph of $f(x)$ passes through point $(5,0)$, so $\lim_{x\to 5} f(x) = 0$.

    (b) 4. At the right-hand edge of the graph, $f(x)$ is straining to reach a height of 4, the same height it intends as x approaches $-\infty$.

Chapter 7

1. Discontinuous. Examining the piecewise-defined function, it's clear that $g(1) = -2$, but using the factoring method, you get $\lim_{x\to 1} g(x) = 5$. Because these are unequal, $g(x)$ is discontinuous at $x = 1$.

2. $a = 12$. Notice that $\lim_{x\to -1^-} h(x) = 6$, because $2(-1)^2 + (-1) - 7 = -6$. (Even though $h(x)$ technically doesn't reach that height, because the domain restriction is $x < -1$, -6 is still the left-hand limit as x approaches -1.) Therefore, $ax + 6 = -6$ when you plug in $x = -1$.

3. $x = 5$ (infinite discontinuity), $x = -5$ (point discontinuity). Factor to get $\frac{(x+5)(2x-5)}{(x+5)(x-5)}$; a limit exists for $x = -5$, but not for $x = 5$.

4. Because $g(1) = -2$ and $g(2) = 4$, we know that all values between -2 and 4 are outputs of g for $1 < x < 2$. Clearly, 0 is between -2 and 4, so the function has a height of 0 (and has an x-intercept) somewhere between $x = 1$ and $x = 2$.

Chapter 8

1. $g'(x) = 10x + 7$; $g'(-1) = -3$. First, calculate $g(x + \Delta x)$:

$$g(x+\Delta x) = 5(x+\Delta x)^2 + 7(x+\Delta x) - 6$$
$$= 5x^2 + 10x\Delta x + 5(\Delta x)^2 + 7x + 7\Delta x - 6$$

After plugging this into the difference quotient and simplifying, you get:

$$\lim_{\Delta x\to 0} \frac{10x\Delta x + 5(\Delta x)^2 + 7\Delta x}{\Delta x}$$

Solve using the factoring method.

$$\lim_{x\to0}\frac{\cancel{x}(10x+5\ x+7)}{\cancel{x}}$$

$$= \lim_{x\to0}\left(10x+5\ x+7\right)$$

$$= 10x+5(0)+7$$

$$= 10x+7$$

2. $\frac{1}{6}$. Begin by calculating $h(8)$:

$$h(x)=\sqrt{x+1}$$

$$h(8)=\sqrt{8+1}$$

$$h(8)=\sqrt{9}$$

$$h(8)=3$$

Apply the alternate difference quotient:

$$\lim_{x\to8}\frac{f(x)-f(8)}{x-8}$$

$$=\lim_{x\to8}\frac{\sqrt{x+1}-3}{x-8}$$

Calculate the limit using the conjugate method:

$$=\lim_{x\to8}\frac{\left(\sqrt{x+1}-3\right)\left(\sqrt{x+1}+3\right)}{\left(x-8\right)\left(\sqrt{x+1}+3\right)}$$

$$=\lim_{x\to8}\frac{x+1-9}{\left(x-8\right)\left(\sqrt{x+1}+3\right)}$$

$$=\lim_{x\to8}\frac{\cancel{x-8}}{\left(\cancel{x-8}\right)\left(\sqrt{x+1}+3\right)}$$

$$=\lim_{x\to8}\frac{1}{\left(\sqrt{x+1}+3\right)}$$

$$=\frac{1}{\sqrt{8+1}+3}$$

$$=\frac{1}{6}$$

Chapter 9

1. (a) $y' = 2x^2 + 6x - 6$. Here is the work behind the scenes:

$$y'=\left(\tfrac{2}{3}\cdot3\right)x^{3-1}+\left(3\cdot2\right)x^{2-1}-\left(6\cdot1\right)x^{1-1}+0$$

(b) $f'(x)=\frac{1}{3x^{2/3}}+\frac{2}{5x^{4/5}}$. Begin by writing the radical terms as fractional exponents and then apply the Power Rule:

$$f(x)=x^{1/3}+2x^{1/5}$$

$$f'(x)=\left(1\cdot\tfrac{1}{3}\right)x^{1/3-1}+\left(2\cdot\tfrac{1}{5}\right)x^{1/5-1}$$

$$=\tfrac{1}{3}x^{-2/3}+\tfrac{2}{5}x^{-4/5}$$

2. To use the Power Rule, you must multiply to get $g(x) = 2x^2 + 7x - 4$ and differentiate that to get $g'(x) = 4x + 7$. Applying the Product Rule gives you:

$$g'(x)=\left(2x-1\right)\left(1\right)+2\left(x+4\right)$$

$$=2x-1+2x+8$$

$$=4x+7$$

3. Make sure to simplify carefully:

$$f'(x)=\frac{(x-5)(12x^3+4x-7)-(3x^4+2x^2-7x)(1)}{(x-5)^2}$$

$$=\frac{(12x^4-60x^3+4x^2-27x+35)-3x^4-2x^2+7x}{x^2-10x+25}$$

$$=\frac{9x^4-60x^3+2x^2-20x+35}{x^2-10x+25}$$

4. $10x(x^2 + 1)^4$. Here you have a function $(x^2 + 1)$ inside another function (x^5). In the Chain Rule formula, $f(x) = x^5$ and $g(x) = x^2 + 1$, because $f(g(x)) = (x^2 + 1)^5$. Therefore, you use the Power Rule to derive the outer function (while leaving $x^2 + 1$ alone) and then multiply by the derivative of $x^2 + 1$ to get $5(x2 + 1)4 \times (2x)$.

5. (a) 19. The instantaneous rate of change is synonymous with the derivative, so find $g'(4)$; according to the Power Rule, $g'(x) = 6x - 5$, so $g'(4) = 19$.

(b) 1. You'll need to find the slope of the secant line, so first get the points representing the x-values of -1 and 3 by plugging those x-values into the equation. Because $g(-1) = 14$ and $g(3) = 18$, the endpoints of the secant line are $(-1,14)$ and $(3,18)$. Calculate the slope of the secant line:

$$10x\left(x^2+1\right)^4$$

6. Begin by writing $\cot x$ as a quotient: $\frac{18-14}{3-(-1)} = \frac{4}{4} = 1$; apply the Quotient Rule to differentiate:

$$\cot x = \frac{\cos x}{\sin x}$$

Factor -1 out of the numerator and use the Mama theorem (see Chapter 4) to replace $\sin^2 x + \cos^2 x$ with 1:

$$\frac{d}{dx}\left(\frac{\cos x}{\sin x}\right) = \frac{\sin x(-\sin x) - \cos x(\cos x)}{\sin^2 x}$$

$$= \frac{-\sin^2 x - \cos^2 x}{\sin^2 x}$$

7. Apply the Quotient Rule to differentiate $k(x)$:

$$= \frac{-\left(\sin^2 x + \cos^2 x\right)}{\sin^2 x}$$

$$= \frac{-1}{\sin^2 x}$$

$$= -\csc^2 x$$

Use the table to identify the function values in the formula and simplify:

$$k'(x) = \frac{g(x) \cdot f'(x) - f(x) \cdot g'(x)}{\left[g(x)\right]^2}$$

$$k'(-1) = \frac{g(-1) \cdot f'(-1) - f(-1) \cdot g'(-1)}{\left[g(-1)\right]^2}$$

8. Consider the following graph of $f(x)$, which includes a tangent line at $x = 4$.

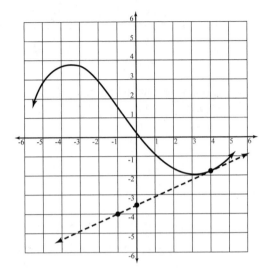

Your tangent line may look slightly different, because this is only an estimate. The tangent line illustrated appears to pass through points $(-1,-4)$ and $(0,-3.5)$. Apply the slope formula:

$$k'(-1) = \frac{2(-6) - (1)(-10)}{2^2}$$

$$= \frac{-12+10}{4}$$

$$= \frac{-2}{4}$$

$$= -\frac{1}{2}$$

Therefore, $\text{slope} = \frac{-3.5-(-4)}{0-(-1)}$

$$= \frac{-3.5+4}{0+1}$$

$$= \frac{0.5}{1}$$

$$= 0.5$$

$$= \frac{1}{2}$$

Chapter 10

1. $y = 15x + 5$. The point of tangency is $(-1,-10)$ and because $g'(x) = 9x^2 - 2x + 4$, $g'(-1) = 15$. Point-slope form gives you $y - (-10) = 15(x - (-1))$, which you can put into slope-intercept form like I did, if you wish.

2. $f'(4) \approx \frac{1}{2}$. The derivative, with respect to x, is $\frac{2}{3}$. Solve this for $4 + x \cdot \frac{dy}{dx} + y - 6y \cdot \frac{dy}{dx} = 0$ to get $\frac{dy}{dx}$. To finish, plug in 3 for x and 2 for y and simplify.

3. 3. Evaluating $f^{-1}(6)$ is the same as solving $\frac{dy}{dx} = \frac{-4-y}{x-6y}$. Square both sides to get $2x^3 - 18 = 36$, and solve for x by adding 18 to both sides, dividing both sides by 2, and then cube rooting both sides of the equation.

4. 0.0945. Remember that $\sqrt{2x^3 - 18} = 6$ and $g^{-1}(2)$ is the solution to the equation $3x^5 + 4x^3 + 2x + 1 = -2$, which is -0.6749465398. Therefore, $\left(g^{-1}\right)'(-2) = \frac{1}{g'\left(g^{-1}(-2)\right)}$.

5. $g^{-1}(-2)$. This is the derivative of the y piece divided by the x piece's derivative. To get the second derivative, derive $\left(g^{-1}\right)'(-2) = \frac{1}{g'(-0.6749465398)} = 0.0945$ (with the Chain Rule) and divide by 2 (the original x equation derivative):
$$\frac{dy}{dx} = \frac{\sec^2 t}{2} = \frac{1}{2}\sec^2 t$$

Chapter 11

1. First of all, $h'(x) = -2x + 6$. When you set that equal to 0 and solve, you get the critical number of $x = 3$. Choose numbers before and after 3 and plug them into the derivative—for example, $h'(2) = 2$ and $h'(4) = -2$. Because the derivative changes from positive to negative, the function changes from increasing to decreasing at $x = 3$, so the critical number represents a relative maximum.

2. Find the derivative: $g'(x) = 6x^2 - 7x - 3$; critical points occur where this equals 0 (it is never undefined). So factor to get $(3x + 1)(2x - 3)$; critical numbers are $\frac{dy}{dx}$ and
$$\frac{d^2y}{dx^2} = \frac{\frac{1}{2} \cdot 2(\sec t) \cdot \sec t \tan t}{2}$$
$$= \frac{\sec^2 \tan t}{2}$$
$$= \frac{1}{2}\sec^2 t \tan t.$$

Pick test values and plug into the derivative to get this wiggle graph:

Because $f'(x)$ is positive on the intervals $x = -\frac{1}{3}$ and $x = \frac{3}{2}$, $f(x)$ is increasing on those intervals.

3. Absolute max: 32; absolute min: −52. Notice that $g'(x) = 3x^2 + 8x + 5$, which factors into $(3x + 5)(x + 1)$, so $\left(-\infty, -\frac{1}{3}\right)$ and $x = -1$ are both critical numbers. A wiggle graph verifies that they are also relative extrema. Test all four x-value candidates, including those and the endpoints: $g(-5) = -52$, $\left(\frac{3}{2}, \infty\right)$, $g(-1) = -4$, and $g(2) = 32$.

4. $x = -\frac{5}{3}$ and $g\left(-\frac{5}{3}\right) \approx -3.852$. If $f(x) = \cos x$, then $f'(x) = -\sin x$ and $f''(x) = -\cos x$. The second derivative wiggle graph for $[0, 2\pi]$ looks like this:

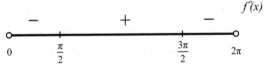

Remember $f(x)$ is concave down wherever $f''(x)$ is negative.

Chapter 12

1. $t = 4$ and $t = 8.196$ seconds. This question is asking, "When is the position equal to −30?" To answer it, use some form of technology to solve the equation $\left(0, \frac{\pi}{2}\right)$. Again, I usually set it equal to 0 and find the x-intercepts (i.e., solve the equation $\left(\frac{3\pi}{2}, 2\pi\right)$). Negative answers make no sense and should be discarded. (Negative time is nonsensical.)

2. The correct order is: the average velocity, the velocity at $t = 7$, and lastly the speed at $t = 3$. The average velocity is the slope connecting the points (2,−4) and (6,−48): $\frac{1}{2}t^3 - 5t^2 + 3t + 6 = -30$ in/sec. The velocity at $t = 7$ is $v(7) = s'(7) = 6.5$ in/sec. The speed at $t = 3$ is the absolute value of the velocity there: $\frac{1}{2}t^3 - 5t^2 + 3t + 36 = 0$ in/sec.

3. $t = 3$ seconds. Because $s''(t) = 3t - 10$, the answer is the solution to the equation $3t - 10 = -1$.

4. (a) $t \approx 10.204$ seconds after launch. Create the position equation: $\frac{-48-(-4)}{6-2} = \frac{-44}{4} = -11$, or $s(t) = -4.9t^2 + 100t + 75$. The highest point reached by the cannonball is the relative maximum of the position equation. Because $\left|s'(3)\right| = \left|-13.5\right| = 13.5$,

then $s(t) = \frac{1}{2}(-9.8)t^2 + 100t + 75$ and $s'(t) = -9.8t + 100$, $0 = -9.8t + 100$, $t = \frac{-100}{-9.8} \approx 10.204$ is the time the ball reaches this height. You can verify that it is a relative maximum using the Second Derivative Test if you like, noting that $s''(t) = -9.8$ is always negative.

(b) 585 meters. In part (a), you determined that the cannonball will reach its maximum height $t = 10.204$ seconds after launch. The position equation gives you the height of the cannonball at any time t seconds after launch. Therefore, the maximum height of the cannonball is $s(10.204)$:

$$s(10.204) = -4.9(10.204)^2 + 100(10.204) + 75$$
$$= -4.9(104.121616) + 1020.4 + 75$$
$$= -510.1959184 + 1020.4 + 75$$
$$\approx 585.204$$

(c) 21.132 seconds. The cannonball will hit the ground when its position is 0—in other words, when $s(t) = 0$. Set the position equation equal to 0 and solve. It is advisable to use your calculator to solve the equation. See the "Technology Focus" section at the end of Chapter 10 for more information.

$$-4.9t^2 + 100t + 75 = 0$$
$$t \approx 21.13245774 \text{ sec}$$

Note that the quadratic has two solutions. The other solution, $t = -0.724$, does not make sense because it is negative. Think about it: You cannot conclude that the cannonball hit the ground 0.724 seconds *before* it launched!

Chapter 13

1. 6. Apply Newton's Method:

$$x_1 = x_0 - \frac{f(x_0)}{f'(x_0)} = 2 - \frac{f(2)}{f'(2)} = 2 - \frac{\sqrt{4}-3}{1/(2\sqrt{4})} = 2 - \frac{-1}{1/4} = 2 + 4 = 6$$

Note that $f(x) = (x+2)^{1/2} - 3$, so

$$f'(x) = \frac{1}{2}(x+2)^{-1/2} \cdot \frac{1}{2(x+2)^{1/2}} = \frac{1}{2\sqrt{x+2}}$$

2. 0. Because x^{-2} has a negative power, move it to the denominator: $\lim\limits_{x\to\infty} \frac{\ln x}{x^2}$. Substitution results in $\frac{\infty}{\infty}$, so apply L'Hôpital's Rule and remember that the derivative of $\ln x$ is $\frac{1}{x}$: $\lim\limits_{x\to\infty} \frac{1/x}{2x}$. This can be rewritten as $\lim\limits_{x\to\infty} \frac{1}{2x^2}$. Substitution now results in 1 divided by a giant number, which is basically 0 according to the third of the special limit theorems from Chapter 6.

3. $x = \frac{1}{2}$. Because $g(x)$ and $g(1) = 1$, the secant slope is $g\left(\frac{1}{4}\right) = 4$. The Power Rule tells you that $\frac{4-1}{1/4-1} = \frac{3}{-3/4} = -4$. Thus, the solution to $g'(x) = -\frac{1}{x^2}$ is the value for c guaranteed by the Mean Value Theorem:

$$-\frac{1}{x^2} = -4$$

Only $x^2 = \frac{1}{4}$

$$\sqrt{x^2} = \pm\sqrt{\frac{1}{4}}$$

$$x = \pm\frac{\sqrt{1}}{\sqrt{4}}$$

$$x = \pm\frac{1}{2}$$

falls in the interval $x = \frac{1}{2}$, so discard the other answer.

4. $\left[\frac{1}{4}, 1\right]$ in²/week. You know from the problem that $\frac{20}{7} \approx 2.857$ if V represents volume. Let's label the surface area S; you want to calculate $\frac{dV}{dt} = 5$. The surface area of a cube

is $S = 6l^2$, where l is the length of a side. Think about it—the surface area of a cube comprises six squares, each having area l^2.

Differentiate the surface area formula to get $\frac{da}{dt}$. You know that $l = 7$, but what is $\frac{da}{dt} = 12s \cdot \frac{ds}{dt}$? To find it, you have to use the given information about $\frac{ds}{dt}$, so you need a second equation containing V.

The volume of a cube with side l is $V = l^3$, so let's derive that baby to get $\frac{dV}{dt} : \frac{dV}{dt}$. You know that $\frac{dV}{dt} = 3s^2 \cdot \frac{ds}{dt}$ and $l = 7$, so plug them into this new equation to get $\frac{dV}{dt} = 5$, so $5 = 3 \cdot 7^2 \cdot \frac{ds}{dt}$.

Now that you finally know what $\frac{ds}{dt} = \frac{5}{147}$ is, plug it back into the $\frac{ds}{dt}$ equation to solve for $\frac{da}{dt}$:

5.
$$\begin{aligned}
\frac{da}{dt} &= 12s \cdot \frac{ds}{dt} \\
&= 12\left(7\right) \cdot \frac{5}{147} \\
&= \frac{420}{147} \\
&= \frac{20}{7} \text{ in}^2/\text{week}
\end{aligned}$$

You want to optimize the product, whose equation is $P = xy$, where x and y are the numbers in question. You know that $y = 2x - 3$, so $P = x(2x - 3) = 2x^2 - 3x$. So $P' = 4x - 3$, and the wiggle graph of P' is

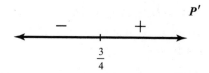

Therefore, one of the numbers is $-\frac{9}{8}$ and the other is $\frac{3}{4}$. Remember, you're asked for the optimal product, so the answer is $y = 2 \cdot \frac{3}{4} - 3 = -\frac{3}{2}$.

Chapter 14

1. The width of all the rectangles will be $xy = \left(\frac{3}{4}\right)\left(-\frac{3}{2}\right) = -\frac{9}{8}$. The left-hand sum will be (you can factor out the $x = \frac{3\pi/2 - \pi/2}{4} = \frac{\pi}{4}$ width from each term to make the answers easier to read): $\frac{\pi}{4}$.

The right-hand sum will be:

$$\frac{\pi}{4}\left[-\cos\frac{\pi}{2} - \cos\frac{3\pi}{4} - \cos\pi - \cos\frac{5\pi}{4}\right]$$

$$= \frac{\pi}{4}\left[0 - \left(-\frac{\sqrt{2}}{2}\right) - (-1) - \left(-\frac{\sqrt{2}}{2}\right)\right]$$

$$= \frac{\pi}{4}\left[\frac{\sqrt{2}}{2} + \frac{\sqrt{2}}{2} + 1\right]$$

$$= \frac{\pi}{4}\left[\frac{2\sqrt{2}}{2} + 1\right]$$

$$= \frac{\pi}{4}\left(\sqrt{2} + 1\right)$$

$$\approx 1.896$$

You'll need a calculator to find the midpoint sum because

$$\frac{\pi}{4}\left(-\cos\frac{3\pi}{4} - \cos\pi - \cos\frac{5\pi}{4} - \cos\frac{3\pi}{2}\right)$$

$$= \frac{\pi}{4}\left[-\left(-\frac{\sqrt{2}}{2}\right) - (-1) - \left(-\frac{\sqrt{2}}{2}\right) - 0\right]$$

$$= \frac{\pi}{4}\left[\frac{\sqrt{2}}{2} + \frac{\sqrt{2}}{2} + 1\right]$$

$$= \frac{\pi}{4}\left[\frac{2\sqrt{2}}{2} + 1\right]$$

$$= \frac{\pi}{4}\left(\sqrt{2} + 1\right)$$

$$\approx 1.896$$

values aren't on the unit circle: $\frac{\pi}{8}$.

2. 1.896. Each trapezoid has width

$$\frac{\pi}{4}\left(-\cos\frac{5\pi}{8} - \cos\frac{7\pi}{8} - \cos\frac{9\pi}{8} - \cos\frac{11\pi}{8}\right) \approx 2.052$$

so according to the Trapezoidal Rule:

$$\Delta x = \frac{\pi - 0}{4} = \frac{\pi}{4}$$

3. 1.622. Each subinterval has the width of

$$\frac{\pi - 0}{2(4)}\left[\sin 0 + 2\sin\frac{\pi}{4} + 2\sin\frac{\pi}{2} + 2\sin\frac{3\pi}{4} + \sin\pi\right]$$

$$= \frac{\pi}{8}\left[0 + 2\left(\frac{\sqrt{2}}{2}\right) + 2(1) + 2\left(\frac{\sqrt{2}}{2}\right) + 0\right]$$

$$= \frac{\pi}{8}\left[\frac{2\sqrt{2}}{2} + \frac{2\sqrt{2}}{2} + 2\right]$$

$$= \frac{\pi}{8}\left[\frac{4\sqrt{2}}{2} + 2\right]$$

$$= \frac{\pi}{8}\left(2\sqrt{2} + 2\right)$$

$$= 1.896$$

Apply the Simpson's Rule formula:
$$\Delta x = \frac{5-1}{4} = 1$$

Chapter 15

1. $\frac{5-1}{3\cdot 4}\left[f(1) + 4f(2) + 2f(3) + 4f(4) + f(5)\right]$

$$= \frac{4}{12}\left[1 + 4\left(\frac{1}{2}\right) + 2\left(\frac{1}{3}\right) + 4\left(\frac{1}{4}\right) + \frac{1}{5}\right]$$

$$= \frac{1}{3}\left[1 + 2 + \frac{2}{3} + 1 + \frac{1}{5}\right]$$

$$= \frac{1}{3}\left[4 + \frac{10}{15} + \frac{3}{15}\right]$$

$$= \frac{1}{3}\left[\frac{60}{15} + \frac{13}{15}\right]$$

$$= \frac{1}{3}\left(\frac{73}{15}\right)$$

$$= \frac{73}{45}$$

$$\approx 1.622$$

Start by writing each term as a separate integral with its own $\frac{2}{5}x^5 + \frac{1}{12}x^4 + \frac{2}{3}x^{3/2} + C$ sign and dx: $\int$. Factor out the coefficients to get $\int 2x^4\,dx + \int \frac{1}{3}x^3\,dx + \int x^{1/2}\,dx$. Apply the Power Rule for integrals and simplify:

$$2\int x^4\,dx + \frac{1}{3}\int x^3\,dx + \int x^{1/2}\,dx$$

Don't get confused when adding 1 to the fractional power:

$$2 \cdot \frac{x^5}{5} + \frac{1}{3} \cdot \frac{x^4}{4} + \frac{x^{3/2}}{3/2} + C$$

$$= \frac{2}{5}x^5 + \frac{1}{12}x^4 + \frac{2}{3}x^{3/2} + C$$

2. $1 + \frac{1}{2} = \frac{3}{2}$. If you write each fraction separately with $-\ln|\cos x| + x + C$ in the denominator of each, you get: $\cos x$.

You memorized the integral of tangent, and the integral of 1 couldn't be easier.

3. -2. The integral of $\cos x$ is $\sin x$ (not $-\sin x$, which is the derivative of $\cos x$). So plug the limits of integration into the integral in the correct order:

$$\int \frac{\sin x}{\cos x}\, dx + \int \frac{\cos x}{\cos x}\, dx$$
$$= \int \tan x\, dx + \int 1\, dx$$

This is the area between the graph of $y = \cos x$ and the x-axis. As you can see in the graph of $\cos x$, the area is below the x-axis, which is why the definite integral is negative.

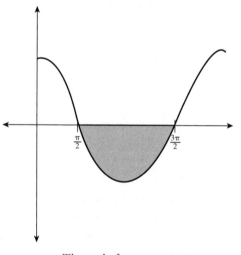

The graph of y = *cos* x.

4. **Part one:** Start by evaluating the definite integral (remember the integral of e^t is e^t):

$$\sin \frac{3\pi}{2} - \sin \frac{\pi}{2}$$
$$= -1 - 1$$
$$= -2$$

Now differentiate; because e is a constant (there is no x exponent) its derivative is 0: $e^{\tan x \times \sec 2}\, x$. (You use the Chain Rule, first leaving the exponent alone and then multiplying by its derivative.)

Part two: Because you are deriving with respect to the variable in the upper bound (and the lower bound is a constant), plug the upper bound into the function and multiply by the upper bound's derivative: $e^{\tan x \times \sec 2}\, x$.

5. $\frac{d}{dx}\left(e^t\Big|_1^{\tan x}\right) = \frac{d}{dx}\left(e^{\tan x} - e\right)$

Set $u = \tan x$ and $du = \sec^2 x\, dx$. Use those two expressions to rewrite the integral using u's: $e^{\tan x} \cdot \sec^2 x$. Don't forget the limits of integration—plug them into $u = \tan x$ to get the new limits: $\tan(0) = 0$ and $\frac{1}{2}$. Integrate $\int u\, du$: $\tan\frac{\pi}{4} = 1$.

Note: you'll get the same final answer if you start with $u = \sec x$ and $du = \sec x \tan x\, dx$.

6. $\int_0^1 u\,du$. *Tricky u-substitution:* set $u = 2x - 3$ and it gives you $\left.\frac{u^2}{2}\right|_0^1 = \frac{1}{2} - 0 = \frac{1}{2}$.

In addition, solve the u equation for x to get $x + 2\ln|2x - 3| + C$. Substitute all of these into the original integral and solve:

$\frac{du}{2} = dx$.

Long division: rewrite $x = \frac{u+3}{2}$ as

$\frac{1}{2}\int \frac{2\left(\frac{u+3}{2}\right)+1}{u}\,du$

$= \frac{1}{2}\int \frac{u+4}{u}\,du$

$= \frac{1}{2}\left(\int du + 4\int \frac{1}{u}\,du\right)$

$= \frac{1}{2}\left(u + 4\ln|u| + C\right)$

$= \frac{1}{2}\left(2x - 3\right) + 2\ln|2x - 3| + C$

$= x + 2\ln|2x - 3| + C$

You get the same answer.

Chapter 16

1. $\frac{2x+1}{2x-3}$. These curves intersect at $x = 0$ and $x = 1$, which you deduce by setting $x^2 = x^3$ and solving for x so those x-values bound the area the functions enclose. The graph of x^2 is above x^3 on that interval, so the area will be $1 + \frac{4}{2x-3}$, which equals: $\frac{1}{12}$.

2. 0.107. According to the Mean Value Theorem for Integration, you know that $\int_0^1 \left(x^2 - x^3\right)dx$. To integrate, you have to use u-substitution with $u = \ln x$ and

$\left(\frac{x^3}{3} - \frac{x^4}{4}\right)_0^1 = \left(\frac{1^3}{3} - \frac{1^4}{4}\right) - 0 = \frac{1}{12}$.

$\left(100 - 1\right)\cdot f\left(c\right) = \int_1^{100} \frac{\ln x}{x}\,dx$

3. 71,000 miles. Distance-traveled problems require you to use the *velocity* equation, so differentiate the position equation to get $v(t) = 3t^2 - 4t - 4$. Now create a wiggle graph of $v(t)$:

$v(t)$

The ship changes direction (i.e., starts heading away from earth) at $t = 2$, so you have to use two integrals for velocity—one for [0,2] and one for [2,5]. Because the integral on [0,2] will be negative, you need to multiply it by -1. Total distance is:

$du = \frac{1}{x}\,dx$

4. (a) $g(4\pi) = 0$. To calculate $g(4\pi)$, plug it into the integral and evaluate it. You'll have to use u-substitution to integrate $\cos 2t$:

$99 f\left(c\right) = \int_{\ln 1}^{\ln 100} u\,du$

$99 f\left(c\right) = \left.\left(\frac{u^2}{2}\right)\right|_{\ln 1}^{\ln 100}$

$99 f\left(c\right) = \frac{\left(\ln 100\right)^2}{2} - 0$

$f\left(c\right) \approx 0.107$

(b) $-\int_0^2 \left(3t^2 - 4t - 4\right)dt + \int_2^5 \left(3t^2 - 4t - 4\right)dt$

$= -\left(-8\right) + 63$

$= 71$

Begin by finding $g'(x)$ using the Fundamental Theorem part two (plug

$g\left(4\pi\right) = \int_{-\pi}^{(4\pi)/2} \cos 2t\,dt$

$= \frac{1}{2}\left.\left(\sin u\right)\right|_{-2\pi}^{4\pi}$

$= 0$

into t and multiply by $g'\left(4\pi\right) = \frac{1}{2}$). Then evaluate the derivative normally: $\frac{\pi}{2}$.

5. $g(x) = x^3$. Use the arc length formula to find the length of each separately: $\frac{1}{2}$.

The cubic graph is steeper, so it covers more ground during the same x-interval.

6. 8.2682. Because

$$g'(x) = \cos\left(2 \cdot \frac{x}{2}\right) \cdot \frac{1}{2}$$

$$g'(x) = \frac{1}{2}\cos x$$

$$g'(4\pi) = \frac{1}{2}\cos(4\pi)$$

$$g'(4\pi) = \frac{1}{2} \cdot 1$$

$$g'(4\pi) = \frac{1}{2}$$

the arc length is:

$$\int_0^2 \sqrt{1 + \left(f'(x)\right)^2}\, dx \qquad \int_0^2 \sqrt{1 + \left(g'(x)\right)^2}\, dx$$

$$= \int_0^2 \sqrt{1 + (2x)^2}\, dx \qquad = \int_0^2 \sqrt{1 + \left(3x^2\right)^2}\, dx$$

$$= \int_0^2 \sqrt{1 + 4x^2}\, dx \qquad = \int_0^2 \sqrt{1 + 9x^4}\, dx$$

$$\approx 4.6468 \qquad\qquad \approx 8.6303$$

Chapter 17

1. $\frac{dx}{dt} = 1$ and $\frac{dy}{dt} = 2t$. Divide both sides by $(x2 - 1)$ and multiply both sides by $\cos y$ to get:

$$\int_1^3 \sqrt{1^2 + (2t)^2}\, dt$$

$$= \int_1^3 \sqrt{1 + 4t^2}\, dt$$

$$\approx 8.268$$

Integrate both sides (use u-substitution for the right side): $y = \arcsin\left(\frac{1}{2}\ln\left|x^2 - 1\right| + C\right)$.

Finally, solve for y by taking the arcsine of both sides (i.e., cancel out sine with its inverse function).

2. Begin by solving the differential equation, separating the x-terms from the y-terms:
$$\cos y\, dy = \frac{x\, dx}{x^2 - 1}.$$

Note that csc y and sin y are reciprocal functions. In other words,
$$\sin y = \frac{1}{2}\ln\left|x^2 - 1\right| + C : \sin y\, dy = x^{4\, dx}.$$

Integrate both sides of the equation:
$$\frac{dy}{\csc y} = x^4 dx.$$

This is the family of solutions to the differential equation. The problem now asks you to find a specific solution whose graph passes through the point $(x,y) = (1,0)$. The easiest way to accomplish this task is to substitute $x = 1$ and $y = 0$ into the equation you just created and solve for C:
$$\frac{1}{\csc y} = \sin y.$$

The specific solution

$$\int \sin y\, dy = \int x^4\, dx$$

$$-\cos y = \frac{x^5}{5} + C$$

$$-\cos(0) = \frac{1^5}{5} + C$$

$$-(1) = \frac{1}{5} + C$$

$$-1 - \frac{1}{5} = C$$

$$-\frac{6}{5} = C$$

passes through the point $(x,y) = (1,0)$.

3. Integrate the acceleration function to find velocity:

$$-\cos y = \frac{x^5}{5} - \frac{6}{5}$$

You know that $v(0) = -2$, so substitute $t = 0$ into $v(t)$ and set the result equal to -2:

$$\left(\text{or} \; -\cos y = \frac{x^5 - 6}{5} \right)$$

Therefore, $v(t) = t^2 + 5t + \cos t - 3$. Integrate to get the position function:

$$v(t) = \int a(t)\,dt$$
$$v(t) = t^2 + 5t + \cos t + C$$

Now use the fact that $s(0) = 5$ to find the corresponding value of C:

$$v(0) = 0^2 + 5(0) + \cos 0 + C$$
$$-2 = 0 + 0 + 1 + C$$
$$-3 = C$$

The final position equation is:

$$s(t) = \int \left(t^2 + 5t + \cos t - 3 \right) dt$$
$$s(t) = \frac{t^3}{3} + \frac{5t^2}{2} + \sin t - 3t + C$$

4. 31.434 days. First things first; you need to calculate k. The initial amount is 15,000, so that will equal N. After $t = 3.82$ days, 7,500 grams remain, so plug into the exponential decay equation:

$$s(0) = \frac{0^3}{3} + \frac{5 \cdot 0^2}{2} + \sin 0 - (3 \cdot 0) + C$$
$$5 = 0 + 0 + 0 - 0 + C$$
$$5 = C$$

Thus, the model for exponential decay is $y = 15{,}000e^{-0.181452t}$. Set it equal to 50 and solve for t to resolve the dilemma:

$$s(t) = \frac{t^3}{3} + \frac{5t^2}{2} + \sin t - 3t + 5$$

Chapter 18

1. 1.08715. The slope of the tangent line to $f(x) = \arctan x$ is:

$$y = Ne^{kt}$$
$$7{,}500 = 15{,}000e^{3.82k}$$
$$\tfrac{1}{2} = e^{3.82k}$$
$$\ln\left(\tfrac{1}{2}\right) = 3.82k$$
$$\frac{\ln(1/2)}{3.82} = k$$
$$-0.181452 \approx k$$

Therefore, the slope of your linear approximation will be:

$$y = 15{,}000e^{-0.181452t}$$
$$50 = 15{,}000e^{-0.181452t}$$
$$\tfrac{1}{300} = e^{-0.181452t}$$
$$\frac{\ln(1/300)}{-0.181452} = t$$

31.4341 days $\approx t$

The point of tangency is (2, arctan 2). That gives the following linear approximation:

$$f'(x) = \frac{1}{1+x^2}.$$

Therefore, arctan 1.9 is approximately equal to: $f'(2) = \frac{1}{1+2^2} = \frac{1}{5}$.

This is pretty close to the actual value: arctan(1.9) $\approx$ 1.08632.

2. The slope field spirals counterclockwise; the specific solution to the differential equation passing through (0,1) should look like the darkened graph:

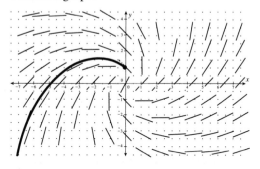

Determine the value of the slopes by plugging into the differential equation. For example, the slope of the segment at point (2,–1) will be:

$$y - \arctan 2 = \tfrac{1}{5}(x - 2)$$

$$y = \tfrac{1}{5}x - \tfrac{2}{5} + \arctan 2$$

3. $\tfrac{1}{5}(1.9) - \tfrac{2}{5} + \arctan 2 \approx 1.08715$. You're traveling a distance of

$$\frac{dy}{dx} = \frac{x+y}{x-y} = \frac{2-1}{2-(-1)} = \frac{1}{3}$$

from point (–1,4), so use that and the given slope to calculate Δy: $\left(-\tfrac{1}{2}, \tfrac{11}{3}\right)$.

So you should go right $\Delta x = \tfrac{1}{2}$ and down

$$m = \frac{y}{x}$$

$$-\frac{2}{3} = \frac{y}{1/2}$$

$$-3 \; y = 2 \cdot \tfrac{1}{2}$$

$$y = -\tfrac{1}{3}$$

from (–1,4) to stay on the line. Make those adjustments to the coordinate to get the answer: $\tfrac{1}{2}$.

4. $-\tfrac{1}{3}$. Here are all three steps:

Step 1: $\left(-1 + \tfrac{1}{2}, 4 - \tfrac{1}{3}\right) = \left(-\tfrac{1}{2}, \tfrac{11}{3}\right)$

knowing $y(1) = \tfrac{16}{27}$, find Δy:

$$\frac{dy}{dx} = 2x - y = 2(0) - 0 = 0$$

This gives you a new point of $x = \tfrac{1}{3}$.

Step 2: $0 = \frac{\Delta y}{1/3}$

$$\Delta y = 0$$

Knowing $\left(0 + \tfrac{1}{3}, 0 + 0\right) = \left(\tfrac{1}{3}, 0\right)$, find Δy:

$$\frac{dy}{dx} = 2x - y = 2\left(\tfrac{1}{3}\right) - 0 = \tfrac{2}{3}$$

This gives you a new point of $\Delta x = \tfrac{1}{3}$.

Step 3: $\tfrac{2}{3} = \frac{y}{1/3}$

$$3 \; y = \tfrac{2}{3}$$

$$y = \tfrac{2}{9}$$

knowing that $\left(\tfrac{1}{3} + \tfrac{1}{3}, 0 + \tfrac{2}{9}\right) = \left(\tfrac{2}{3}, \tfrac{2}{9}\right)$, find Δy:

$$\frac{dy}{dx} = 2x - y = 2\left(\tfrac{2}{3}\right) - \tfrac{2}{9} = \tfrac{10}{9}$$

This gives you a new point of $\Delta x = \tfrac{1}{3}$.

Glossary

absolute extrema point   The highest or lowest point on a graph.

acceleration   The rate of change of velocity.

accumulation function   A function defined by a definite integral; it has a variable in one or both of its limits of integration.

antiderivative   The opposite of the derivative; if $f(x)$ is an antiderivative of $g(x)$, then $\int g(x)dx = f(x)+C$, where C is a constant.

antidifferentiation   The process of creating an antiderivative or integral.

asymptote   A line representing an unattainable value that shapes a graph; because the graph cannot achieve the value, the graph bends toward that line but won't intersect it.

average value of a function   The value, $f(c)$, guaranteed by the Mean Value Theorem for Integration found via the equation $f(c) = \frac{\int_a^b f(x)dx}{b-a}$.

Chain Rule   The derivative of the composite function $h(x) = f(g(x))$ is $h'(x) = f'(g(x)) \cdot g'(x)$.

cofunction   Trigonometric functions with the same name, apart from the prefix "co-," like sine and cosine or tangent and cotangent.

concavity   Describes how a curve bends; a curve that can hold water poured into it from the top of the graph is concave up, whereas one that cannot hold water is concave down.

conjugate   A binomial whose middle sign is the opposite of another binomial with the same terms (e.g., $3+\sqrt{x}$ and $3-\sqrt{x}$ are conjugates).

constant   A polynomial of degree 0; a real number.

constant of integration   The unknown constant that results from an indefinite integral, usually written as C in your solution; it is a required piece of all indefinite integral solutions.

continuous   A function $f(x)$ is continuous at $x = c$ if $\lim_{x \to c} f(x) = f(c)$.

coterminal angles   Angles that have the same function value, because the space between them is a multiple of the function's period.

critical number   An x-value that causes a function to equal zero or become undefined.

cubic   A polynomial of degree 3.

definite integral   An integral that contains limits of integration; its solution is a real number.

degree   The largest exponent in a polynomial.

derivative   The derivative of a function $f(x)$ at $x = c$ is the slope of the tangent line to f at $x = c$, usually written $f'(c)$.

difference quotient   One of two formulas that define a derivative:

$$f'(x) = \lim_{x \to 0} \frac{f(x+x)-f(x)}{x} \qquad \text{or} \qquad f'(c) = \lim_{x \to c} \frac{f(x)-f(c)}{x-c}.$$

differentiable   Possessing a derivative at the specific x-value; if a function does not have a derivative at the given x-value, it is said to be "nondifferentiable" there.

differential equation   An equation containing a derivative.

displacement   The total change in position counting only the beginning and ending position; if the object in question changes direction any time during that interval of time, it does not correctly reflect the total distance traveled.

domain   The set of possible inputs for a function.

essential discontinuity   *See* infinite discontinuity.

Euler's Method   A technique used to approximate solutions to a differential equation when you can't apply separation of variables.

everywhere continuous   A function that is continuous at every x in its domain.

exponential growth and decay   A population grows or decays exponentially if its rate of change is proportional to the population itself—in other words, $\frac{dP}{dt} = k \cdot P$, where k is a constant and P is the size of the population.

extrema point   A high or low point in the curve, a *maximum* or a *minimum*, respectively; it represents an extreme value of the graph, whether extremely high or extremely low, in relation to the points surrounding it.

Extreme Value Theorem   If a function $f(x)$ is continuous on the closed interval $[a,b]$, then $f(x)$ has an absolute maximum and an absolute minimum on $[a,b]$.

factoring   Reversing the process of multiplication. The results of the factoring process can be multiplied together to get the original quantity.

family of solutions   Any mathematical solution containing "+ C"; it compactly represents an infinite number of possible solutions, each differing only by a constant.

function   A relation such that every input has exactly one matching output.

greatest common factor   The largest quantity by which all the terms of an expression can be divided evenly.

implicit differentiation   Allows you to find the slope of a tangent line when the equation in question cannot be solved for y.

indefinite integral   An integral that does not contain limits of integration; its solution is the antiderivative of the expression (and must contain a constant of integration).

indeterminate form   An expression whose value is unclear; the most common indeterminate forms are $\pm\frac{\infty}{\infty}$, $\pm\frac{0}{0}$, and $0 \cdot \infty$.

infinite discontinuity   Discontinuity caused by a vertical asymptote. Also called *essential discontinuity*.

inflection points   Points on a graph where the concavity changes.

integer   A number without a decimal or fractional part.

integral   The opposite of the derivative; if $f(x)$ is the integral of $g(x)$, then $\int g(x)\,dx = f(x) + C$, where C is a constant.

integration   The process of creating an antiderivative or integral.

intercept   Numeric value at which a graph hits either the x- or y-axis.

Intermediate Value Theorem   If a function $f(x)$ is continuous on the closed interval $[a,b]$, then for every real number d between $f(a)$ and $f(b)$, there exists a c between a and b so that $f(c) = d$.

irrational root   An x-intercept that cannot be written as a fraction.

jump discontinuity   Occurs when no general limit exists at the given x-value because the left- and right-hand limits are not equal.

left sum   A Riemann approximation in which the heights of the rectangles are defined by the values of the function at the left-hand side of each interval.

left-hand limit   The height a function intends to reach as you approach the given x-value *from* the left.

L'Hôpital's Rule   If a limit results in an indeterminate form after substitution, you can take the derivatives of the numerator and denominator of the fraction separately without changing the limit's value $\lim_{x \to c} \frac{f(x)}{g(x)} = \lim_{x \to c} \frac{f'(x)}{g'(x)}$.

limit   The height a function *intends* to reach at a given *x*-value, whether or not it actually reaches it.

limits of integration   Small numbers next to the integral sign, indicating the boundaries when calculating area under the curve; in the expression $\int_1^3 x^5 \, dx$, the limits of integration are 1 and 3.

linear approximation   The equation of a tangent line to a function used to help approximate the function's values lying close to the point of tangency.

linear expression   A polynomial of degree 1.

logistic growth   Begins quickly (it initially looks like exponential growth) but eventually slows and levels off to some limiting value; most natural phenomena, including population and sales graphs, follow this pattern rather than exponential growth.

Mean Value Theorem   If a function $f(x)$ is continuous and differentiable on a closed interval $[a,b]$, then there exists a point c, $a \leq c \leq b$, so that $f'(c) = \frac{f(b) - f(a)}{b - a}$.

Mean Value Theorem for Integration   If a function $f(x)$ is continuous on the interval $[a,b]$, then there exists a c, $a \leq c \leq b$, such that $(b - a) \cdot f(c) = \int_a^b f(x) \, dx$.

midpoint sum   A Riemann approximation in which the heights of the rectangles are defined by the values of the function at the midpoint of each interval.

nondifferentiable   Not possessing a derivative.

nonremovable discontinuity   A point of discontinuity for which no limit exists (e.g., infinite or jump discontinuity).

normal line   The line perpendicular to a function's tangent line at the point of tangency.

optimizing   Finding the maximum or minimum value of a function given a set of circumstances.

parameter   A variable into which you plug numeric values to find coordinates on a parametric equation graph.

parametric equations   Pairs of equations, usually in the form of "*x* =" and "*y* =," that define points of a graph in terms of yet another variable, usually *t* or *θ*.

period   The amount of horizontal space it takes a periodic function to repeat itself.

periodic function   A function whose values repeat over and over after a fixed interval.

point discontinuity   Occurs when a general limit exists but the function value is not defined.

point-slope form   A line containing the point (x_1,y_1) with slope m has equation $y - y_1 = m(x - x_1)$.

position equation   A mathematical model that outputs an object's position at a given time, t.

Power Rule for Differentiation   The derivative of the expression ax^n with respect to x, where a and n are real numbers, is $(a \cdot n)x^{n-1}$.

Power Rule for Integration   The integral of a single variable to a real-number power is found by adding 1 to the existing exponent and dividing the entire expression by the new exponent $\int x^n \, dx = \frac{x^{n+1}}{n+1} + C$, assuming $n \neq -1$.

Product Rule   The derivative of $f(x)g(x)$, with respect to x, is $f(x) \cdot g'(x) + f'(x) \cdot g(x)$.

quadratic   A polynomial of degree 2.

Quotient Rule   If $h(x) = \frac{f(x)}{g(x)}$, then $h'(x) = \frac{g(x) \cdot f'(x) - f(x) \cdot g'(x)}{\left[g(x)\right]^2}$.

range   The set of possible outputs for a function.

reciprocal   The fraction with its numerator and denominator reversed (e.g., the reciprocal of $\frac{7}{4}$ is $\frac{4}{7}$).

related rates   A problem that uses a known rate of change to compute the rate of change for another variable in the problem.

relation   A collection of related numbers, usually described by an equation.

relative extrema point   Occurs when that point is higher or lower than all of the points in the immediate surrounding area; visually, a relative maximum is the peak of a hill in the graph, and a relative minimum is the lowest point of a dip in the graph.

removable discontinuity   A point of discontinuity for which a limit exists (i.e., point discontinuity).

Riemann sum   An approximation for the area beneath a curve calculated by adding the areas of rectangles.

right sum   A Riemann approximation in which the heights of the rectangles are defined by the values of the function at the right-hand side of each interval.

right-hand limit   A function's intended height as you approach the given x-value *from* the right.

Rolle's Theorem   If a function $f(x)$ is continuous and differentiable on a closed interval $[a,b]$ and $f(a) = f(b)$, then there exists a c between a and b such that $f'(c) = 0$.

secant line   A line that cuts through a graph, usually intersecting it in multiple locations.

separation of variables   A technique used to solve basic differential equations; in it, you move the different variables of the equation to different sides of the equal sign in order to integrate each side of the equation separately.

sign graph   *See* wiggle graph.

Simpson's Rule   The approximate area under the curve $f(x)$ on the closed interval $[a,b]$ using an even number of subintervals, n, is: $\frac{b-a}{3n}\left[f(a)+4(x_1)+2f(x_2)+\cdots+2f(x_{n-2})+4f(x_{n-1})+f(b)\right]$.

slope   Numeric value that describes the "slantiness" of a line.

slope field   A tool to visualize the solution of a differential equation; a collection of line segments centered at points whose slopes are the values of the differential equation evaluated at those points.

slope-intercept form   A line with slope m and y-intercept b has equation $y = mx + b$.

speed   The absolute value of velocity.

symmetric function   A function that looks like a mirror image of itself, typically across the x-axis, y-axis, or about the origin. Symmetry across the x-axis is possible as well but results in a graph that is not a function.

tangent line   A line that skims across a curve, hitting it only once at the indicated location.

Trapezoidal Rule   The approximate area beneath a curve $f(x)$ on the interval $[a,b]$ using n trapezoids is: $\frac{b-a}{2n}\left[f(a)+2f(x_1)+2f(x_2)+2f(x_3)+\cdots+2f(x_{n-1})+f(b)\right]$.

***u*-substitution**   Integration technique that is useful when a function and its derivative appear in an integral.

velocity   The rate of change of position; it includes a component of direction, and therefore may be negative.

vertical line test   Tests whether or not a graph is a function; if any vertical line can be drawn through the graph that intersects the graph more than once, then the graph in question cannot be a function.

wiggle graph   A segmented number line that describes the direction of a function and the signs of its derivative.

Index

D

E

F

I

Q

T